Cellular Organelles

PRINCIPLES OF MEDICAL BIOLOGY
A Multi-Volume Work, Volume 2

Editors: **E. EDWARD BITTAR,** *Department of Physiology, University of Wisconsin*

NEVILLE BITTAR, *Department of Medicine, University of Wisconsin, Madison*

Principles of Medical Biology
A Multi-Volume Work

Edited by **E. Edward Bittar,** *Department of Physiology, University of Wisconsin, Madison* and
Neville Bittar, *Department of Medicine, University of Wisconsin, Madison*

This work in 25 modules provides:

- A holistic treatment of the main medical disciplines. The basic sciences including most of the achievements in cell and molecular biology have been blended with pathology and clinical medicine. Thus, a special feature is that departmental barriers have been overcome.
- The subject matter covered in preclinical and clinical courses has been reduced by almost one-third without sacrificing any of the essentials of a sound medical education. This information base thus represents an integrated core curriculum.
- The movement toward reform in medical teaching calls for the adoption of an integrated core curriculum involving small-group teaching and the recognition of the student as an active learner.
- There are increasing indications that the traditional education system in which the teacher plays the role of expert and the student that of a passive learner is undergoing reform in many medical schools. The trend can only grow.
- Medical biology as the new profession has the power to simplify the problem of reductionism.
- Over 700 internationally acclaimed medical scientists, pathologists, clinical investigators, clinicians and bioethicists are participants in this undertaking.

The next seven physical volumes are planned for Fall 1996.

Cellular Organelles

Edited by **E. EDWARD BITTAR**
Department of Physiology
University of Wisconsin
Madison, Wisconsin

NEVILLE BITTAR
Department of Medicine
University of Wisconsin
Madison, Wisconsin

JAI PRESS INC.

Greenwich, Connecticut *London, England*

Library of Congress Cataloging-in-Publication Data

Cellular organelles / edited by E. Bittar, Neville Bittar.
p. cm.—(Principles of medical biology; v.2)
Includes index.
ISBN 1-155938-803-X.
1. Cell organelles. I. Bittar, E. Edward. II. Bittar, Neville.
III. Series.
[DNLM: 1. Organelles—[physiology. QH 591 C393 1995]
QH581.2.C45 1995
574.87'34—dc20
DNLM/DLC
for Library of Congress 95-16568
CIP

55 Old Post Road, No 2
Greenwich, Connecticut 06836

JAI PRESS LTD.
The Courtyard
28 High Street
Hampton Hill
Middlesex TW12 1PD
England

ISBN: 1-55938-803-X

Library of Congress Catalog Number: 95-16568
Manufactured in the United States of America

Transferred to digital printing 2008

CONTENTS

LIST OF CONTRIBUTORS

Richard Brimacombe	Max-Planck-Institut für Molekulare Genetik Berlin-Dahlem, Germany
A. DeSilva	Department of Biology University of Rochester Rochester, New York
David S. Ettenson	Centre for Cardiovascular Research University of Toronto Toronto, Ontario, Canada
Robert M. Evans	Department of Pathology University of Colorado Health Sciences Center Boulder, Colorado
Paul A. Fisher	Department of Pharmacological Sciences Health Sciences Center State University of New York Stony Brook, New York
David Goldfarb	Department of Biology University of Rochester Rochester, New York
Avrum I. Gotlieb	Centre for Cardiovascular Research University of Toronto Toronto, Ontario, Canada
Daniele Hernandez-Verdun	Institut Jacques Monod Paris, France
Henriette R. Junera	Institut Jacques Monod Paris, France
Michael W. Klymkowsky	Molecular, Cellular, and Developmental Biology University of Colorado Boulder, Colorado

Gordon L.E. Koch	Medical Research Council Laboratory of Molecular Biology Cambridge, England
S.K. Malhotra	Department of Zoology University of Alberta Edmonton, Alberta, Canada
Anthony N. Martonosi	Department of Biochemistry and Molecular Biology State University of New York Health Science Center Syracuse, New York
N. Pokrywka	Department of Biology Vassar College Poughkeepsie, New York
J.B. Rattner	Departments of Anatomy and Medical Biochemistry The University of Calgary Calgary, Alberta, Canada
T.K. Shnitka	Department of Pathology University of Alberta Edmonton, Alberta, Canada
Nico Stuurman	Department of Pharmacological Sciences Health Sciences Center State University of New York Stony Brook, New York
M. Zillmann	Department of Biology University of Rochester Rochester, New York

PREFACE

The purpose of this volume is to provide a synopsis of present knowledge of the structure, organization, and function of cellular organelles with an emphasis on the examination of important but unsolved problems, and the directions in which molecular and cell biology are moving. Though designed primarily to meet the needs of the first-year medical student, particularly in schools where the traditional curriculum has been partly or wholly replaced by a multi-disciplinary core curriculum, the mass of information made available here should prove useful to students of biochemistry, physiology, biology, bioengineering, dentistry, and nursing.

It is not yet possible to give a complete account of the relations between the organelles of two compartments and of the mechanisms by which some degree of order is maintained in the cell as a whole. However, a new breed of scientists, known as molecular cell biologists, have already contributed in some measure to our understanding of several biological phenomena notably inter-organelle communication. Take, for example, intracellular membrane transport: it can now be expressed in terms of the sorting, targeting, and transport of protein from the endoplasmic reticulum to another compartment.

This volume contains the first ten chapters on the subject of organelles. The remaining four are in Volume 3, to which sections on organelle disorders and the extracellular matrix have been added.

We would like to take this opportunity of thanking the contributing authors for their enthusiasm, cooperation, and forbearance. We also wish to thank members of the editorial and production staff of JAI Press for their assistance and courtesy.

E. EDWARD BITTAR
NEVILLE BITTAR

Chapter 1

The Plasma Membrane:
Membrane Proteins and their Interactions

S.K. MALHOTRA and T.K. SHNITKA

INTRODUCTION

The plasma membrane constitutes the essential physiological barrier at the surface of cells (Nageli, 1855). Also, cells have a different internal environment from the milieu which surrounds them. This difference is maintained by the plasma membrane which is responsible for ion and fluid transport, the absorption of small molecules (e.g., glucose and

Principles of Medical Biology, Volume 2
Cellular Organelles, pages 1-18

ISBN:1-55938-803-X

amino acids) and the uptake of macromolecules and particulate material by endocytosis. Plasma membranes also interact with surrounding extra-cellular matrix molecules (ECMS) and with the underlying intracellular cytoskeletal framework and its associated proteins, which regulate cell shape and respond to external and internal stimuli. The plasma membrane is made up of a lipid bilayer (Gorter and Grendel, 1925) and associated membrane proteins. Notwithstanding, the molecular diversity of lipids (and glycolipids) and proteins (and glycoproteins) inherent in the plasma membranes of different cell types, all share many major characteristics in common. The dynamic nature of the plasma membrane is depicted in the fluid mosaic model (Singer and Nicolson, 1972; Singer, 1992; see Malhotra, 1983), which is envisaged as a two-dimensional solution of proteins in a viscous lipid bilayer. Both the lipids and proteins are distributed asymmetrically in the two halves of the bilayer and are capable of varying degrees of mobility in the plane of the membrane. The proteins are either completely or partially intercalated into the hydrophobic domain of the lipid bilayer (integral membrane proteins) or attached to the surface of the lipid bilayer, largely by electrostatic interactions. The latter can be readily removed by changing the pH or the ionic strength of buffers employed in extraction procedures. Many of the membrane proteins show covalently linked lipids post-translationally added to the polypeptides, either bound to myristic acid (e.g., α-subunit of GTP binding protein), or to palmitic acid (e.g., transferrin receptor), or to glycosyl- phosphatidylinositol (e.g., neural cell adhesion molecules (N-CAM). Carbohydrates are also attached to membrane proteins, either through N-linked glycosylation or O-linked glycosylation which occurs in the ER and Golgi apparatus (see Gennis, 1989).

The free surfaces of cells are covered by an external coat of variable thickness, which has been termed the *glycocalyx* in animal cells, on the basis of morphological findings discernible by carbohydrate histochemistry for light microscopy and by electron microscopy (see Rambourg and Leblond, 1967; Ito, 1974). Great diversity in the glycoproteins on the surface of the plasma membrane is evident from the presence therein of a large variety of receptors, transport proteins, enzymes and cell adhesion molecules. The function of cell surface receptors is to recognize specific ligands from among a myriad of extracellular stimuli, and to then activate an effector system that sets in motion a complex intracellular second messenger cascade.

The research conducted on the plasma membrane over the past three decades is far too extensive to be treated in its entirety in a single chapter. For this reason, we are limiting our presentation to general remarks concerning recent work that has been done on the integral proteins of the plasma membrane, with emphasis on the molecular assemblies and mechanisms that subserve certain defined plasma membrane functions. For a comprehensive general survey of the molecular biology of the plasma membrane, students should consult one of the several excellent textbooks now available.

MEMBRANE PROTEINS

Despite the diverse and specific roles served by the plasma membranes of various differentiated cell types, there is a basic similarity in the three-dimensional structure of different integral membrane proteins, whether they serve the role of receptors, form ion channels, or catalyze enzymatic reactions. High resolution electron microscopy (7-9 Å) of two-dimensional lattices within the membrane coupled with three-dimensional image reconstructions have revealed predominantly (α-helical conformations that traverse the membranes (see Unwin and Henderson, 1984). Each α-helical rod is made up of at least 22-amino acid (aa) residues sufficiently hydrophobic to traverse the hydrophobic lipid bilayer (Lodish, 1988). Stretches of amino acids capable of forming an a-helix likely to span the hydrophobic core of the lipid bilayer can be determined from the amino acid sequence (hydropathy profiles) and the helix may be largely composed of hydrophobic amino acids, or the helix may be amphipathic with the nonpolar face towards the fatty acid chains of the lipid bilayer. The bacteriorhodopsin molecule of the purple membranes of the photosynthetic bacterium *Holobacterium* has seven a-helices spanning the lipid bilayer (Unwin and Henderson, 1984). Each of the five subunits of the nicotinic acetylcholine (Ach) receptor molecule has five helices (see Unwin, 1993), while the band 3 protein (MW 100 kD) which forms the dimeric channel for exchange of Cl and HCO_3 in red blood cells has 12 transmembrane segments. Connexin, the only known protein of the gap junction, has 4 transmembrane segments in each of the two sets of identical subunits that make up a connexon (Kumar and Gilula, 1992). Na^+- K^+- , and Ca^{2+}-channel-forming polypeptides, contain six transmembrane segments in each of their polypeptides. They all have similar features in forming the pore through the membrane which reflects a common architectural plan. This similarity in design is evident in the pore-forming peptides irrespective of the pore size and the number of subunits forming the pore. The pore in the gap junction is ~16 Å, the acetylcholine receptor channel is ~7Å, and the Na^+ channel is ~4 Å (Unwin, 1986); the number of subunits contributing to the formation of the pore is respectively six, five, or four. Both the gap junction channel and the Ach channel are much wider at one end and become narrower at the other end.

The structure of the nicotinic Ach receptor channel provides one of the best examples of a high resolution image (up to 9 Å) obtained from three-dimensional lattices of pore-forming membrane proteins (Unwin, 1993). Unwin has presented models incorporating the data from the amino acid sequences which suggest that the side chains of the highly conserved leucines (position 251) in the five subunits contributing to the formation of a pore are likely to influence ion transport through the pore. It is of physiological interest that the leucines in this position are also conserved in the subunits of GABA, glycine, and serotonin receptor molecules (see Unwin, 1993). Unwin's model also depicts the distribution of various amino acids along the length of the cation selective pore.

Beside α-helices, it now appears from the three-dimensional image reconstructions of Ach receptors, that β-sheets also may contribute to the three dimensional structure of membrane proteins. For example, β-sheets are predominant in the aqueous channels formed by the porins in the outer membrane of Gram negative bacteria (Cowan et al., 1992). Porins are predominantly polar and contain no long stretches of hydrophobic segments (which are the likely candidates for traversing the hydrophobic lipid bilayer). The porins form relatively large aqueous channels through which ions and small organic molecules pass into the periplasmic space; from there they traverse the plasma membrane through specific active transport proteins. Porins also form channels in the outer membranes of mitochondria and chloroplasts, and allow the passage of large molecules (~up to 6000 MW in mitochondria, and 13,000 MW in chloroplasts), but exclude macromolecules and enzymes (see Wolfe, 1993).

The best example of a high resolution structure of an integral membrane protein is the photosynthetic reaction center in the plasma membrane of the sulfur bacterium *Rhodopseudomonas viridis* (Deisenhofer and Michel, 1989). This protein has four polypeptides, three of which span the membrane, two of which (referred to as L-and M-subunits) have five transmembrane α- helices, each of which is arranged in a crescent; the third subunit (H) is anchored to the cytoplasmic face of the membrane by a single transmembrane α-helix. The fourth subunit, which is a cytochrome, is a peripheral membrane protein bound to the other three subunits on the exoplasmic face and this lies in the exoplasmic space. (For the elucidation of the molecular structure of this crystallized reaction center, Deisenhofer, Michel, and Huber were awarded the Nobel Prize in Chemistry in 1988.)

In general, the overall structure of integral membrane proteins is conserved, although the amino acids themselves may not be. The side chains that extend outwards and anchor proteins in the membrane are hydrophobic, but the side chains that face inwards and bind helices together interact through van der Waals forces (Darnell et al., 1990).

In recent years valuable information has been gathered on the involvement of plasma membrane constituents in recognizing specific extracellular signals and then activating appropriate cellular responses through the generation of intracellular second messengers, namely cAMP, cGMP, 1,2 diacylglycerol (1,2DG), and inositol 1,4,5- triphosphate (IP_3). Also, the role has been determined of tyrosine autophosphorylation of receptors (i.e., in receptors for insulin and growth factors) (Pazin and Williams, 1992). The discovery of domains for binding sites for specific tyrosine phosphoproteins (*src* homology 2, SH2, ~ 100 aa) and specific proline motifs (SH3, ~ 60 aa), constitute important advances in our understanding of the mechanisms of signal transduction during cellular responses to ligands acting upon receptors at the plasma membrane (see McCormick, 1993). Another recent discovery is the finding of a tyrosine kinase at focal adhesion sites in chicken embryo fibroblasts, detected by antibody staining. This kinase has been termed $pl25^{FAK}$ (predicted molecular mass of 116 kD), and it has been suggested that this kinase represents a convergence in the action of diverse groups of molecules that have similar indirect effects on cellular responses, such as are generated by the activation of integrins involved in the adhesion of cells to extracellular proteins and the transmission of signals into cells (Hynes, 1992), in oncogenic forms of pp 60 *src,* and in mitogenic neuropeptides (Zachary and Rozengurt, 1992). $pl25^{FAK}$ contains a central catalytic domain that is characteristic of the catalytic domain of other protein tyrosine kinases, but differs from them in lacking their noncatalytic motifs (Zachary and Rozengurt, 1992). $pl25^{FAK}$ also appears to have no membrane association sites or the SH2 and SH3 domains of tyrosine phosphoproteins. The precise mechanism for the functioning of $p125^{FAK}$ in cellular regulation in response to activation of diverse receptors is not yet known (Zachary and Rozengurt, 1992).

ENDOCYTOSIS

The plasma membrane is actively involved in the uptake by endocytosis of a variety of materials from the surrounding medium. The most extensively studied aspect of endocytosis is receptor-mediated endocytosis which takes place at small specialized sites which are readily recognized as coated pits by a bristle-like thickening on the cytoplasmic surface of the plasma membrane (Figure 1). The membrane coat-material is the protein clathrin, which forms a complex network of pentagons and hexagons on the surface of the plasma membrane. After the coated pits are invaginated and pinched off, the clathrin is removed from the vesicle by an ATPase which belongs to the family of 70-kD heat-shock proteins. The fate of the receptors and the bound ligands in the endocytosed vesicles is conditioned by the nature of the materials taken up by the cell. The best studied example is the endocytosis of low-density lipoprotein (LDL) particles, which carry cholesterol into liver cells and other types of cells. Brown and Goldstein received the Nobel Prize in 1985 for their elucidation of the structure of the receptor for LDL, the detailed analysis of the pathway followed by endocytosed LDL particles, and the regulation and utilization of cholesterol in cells. They and their colleagues also elucidated mutations in the structure of the LDL receptor that result in cholesterol-related diseases, such as familial hypercholesterolemia (see Motulsky, 1986).

The uptake of LDL in coated vesicles is similar to that of several other substances which enter cells via receptor-mediated endocytosis, e.g., transferrin, epidermal growth factor, and some lysosomal proteins. pH changes play an important role in the separation of the respective receptors from their ligands and their ultimate fate in vesicular traffic through the cell.

While the role of the clathrin-coated pits of the plasma membrane in the endocytosis of receptor-mediated ligands is well documented, the role of non-clathrin- coated regions of the plasma membrane in endocytosis is not as well established. It appears that cells may make use of the latter phenomenon for the uptake and recycling of certain cell surface molecules, such as the major histocompatibility glycoproteins resident on human lymphoblastoid cells (see Watts and March, 1992).

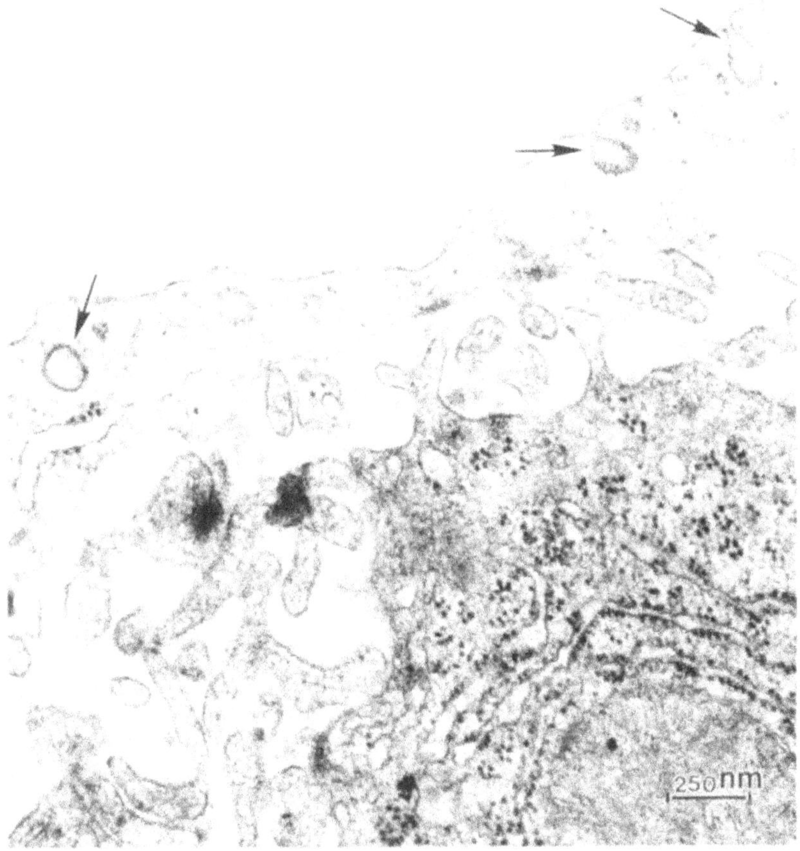

Figure 1. Electron micrograph showing "coated" vesicles (*arrows*) in a Kupffer cell in a thin section of rat liver.

MEMBRANE FUSION

The fusion of membranes in endocytosis and exocytosis is dependent upon interactions between membrane proteins and an intermixing of membrane lipids (White, 1992). Annexins, which constitute a family of proteins containing highly conserved Ca^{2+}-binding repeats (4 in proteins of 35-37 kD, and 8 in proteins of 65 kD) are thought to facilitate close contacts between membranes destined to undergo fusion. There are at least 13 annexins known thus far, and the precise mechanisms(s) responsible for membrane fusion are still being investigated. Membrane fusion is a widespread phenomenon, and apart from annexins, recognition receptors and well-conserved fusion peptides contribute to the membrane fusion process (see White, 1992). It is likely that the membrane fusion process has evolved on a common theme in biology, and that the similarity in the overall properties of fusion peptides in enveloped viruses with those in the sperm head indicates highly conserved motifs. The fusion peptides are short, relatively hydrophobic segments, consisting of 16 to 26 amino acids. They are a part of the membrane-anchored subunit of the protein, and can be located at the N-terminal or internally in the polypeptide (Aitken, 1992; Blobel et al., 1992; White, 1992).

PLASMA MEMBRANE AND CYTOSKELETAL INTERACTIONS

In addition to the role of annexins in membrane-to-membrane interactions, annexins may be involved in the binding of actin fibers to the plasma membrane. Actin and other cytoskeletal proteins are implicated in many membrane-related processes, such as cytokinesis, phagocytosis and plasma membrane ruffling (see Gruenberg and Emans, 1993). Interactions between the plasma membrane proteins and the underlying cytoskeletal and other associated proteins provide the basis for cell motility, the generation and maintenance of cell shape, cell surface domains, cell polarity, and form a communication link between the cell surface and the nucleus (see Nelson, 1992; Figure 2).

The regulation of actin polymerization/depolymerization is important in the process of cell movement. Shariff and Luna (1992) have shown that diacylglycerol (DG) stimulates actin polymerization in the presence of highy purified plasma membranes from the slime mold *Dictyostelium discoideum*. Polymerization takes place in the presence of an as yet unknown peripheral protein, which may contain a DG binding site, and thereby promote the formation of actin nuclei. Thus, DG generated during the phospholipase-catalyzed hydrolysis of phospholipids, particularly phosphatidylinositide (PI) and other components of signal transduction pathways, could control the assembly of actin at the inner face of the plasma membrane.

Another variety of interaction of plasma membrane proteins with cytoskeletal proteins has been well documented in skeletal muscle and in the electric organ of the *Torpedo*. Such interactions are particularly evident in the region of the postsynaptic membrane, where the Ach receptors cluster (see Froehner, 1991). A peripheral membrane protein

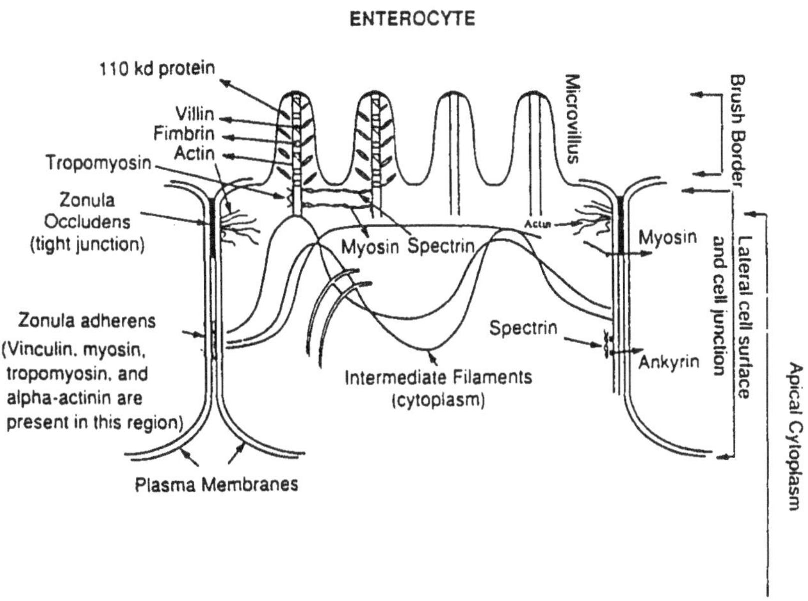

Figure 2. Diagrammatic representation of the cytoskeleton in the apical region of the intestinal epithelial cell (enterocyte). This diagram is constructed from data previously published; for example, see Mooseker 1985). Reproduced from Malhotra and Shnitka (1991).

(43-kD) has been shown to induce clustering of Ach receptors in fibroblasts transfected with the receptor subunits and an expression construct encoding the 43-kD protein. Such an interaction between the receptor molecules and the 43-kD protein provides tentative evidence for a direct contact between the two proteins (Phillips et al., 1991). Other proteins (58 and 87 kD) concentrated in the postsynaptic region underneath the plasma membrane are implicated in anchoring the receptors in the plasma membrane. The 87-kD protein is a tyrosine kinase which has homologous domains (cysteine-rich and C terminal) with the now well-known skeletal muscle protein dystrophin (427 kD). Dystrophin is concentrated in the postsynaptic region but is also associated with the nonsynaptic plasma membrane. It is also present in brain (Nudel et al., 1989). Dystrophin belongs to the actin binding protein, spectrin superfamily, and it is thought to bind to a specific glycoprotein complex in the membrane. The 87-kD protein forms a complex with the 58-kD protein

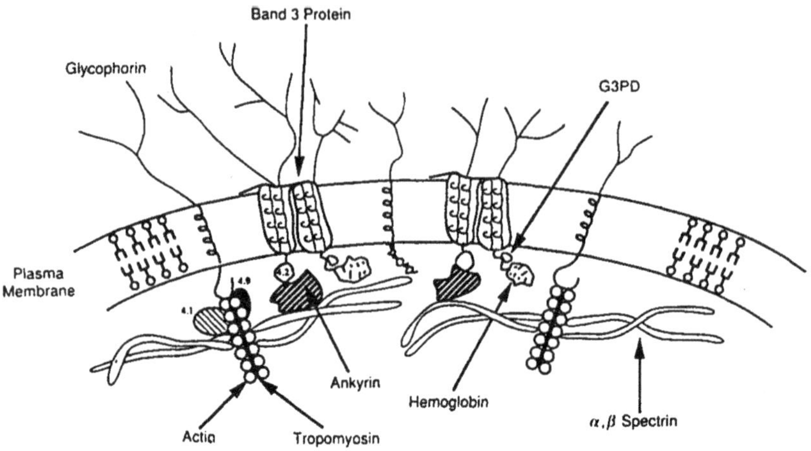

Figure 3. Diagrammatic representation of the red blood cell cytoskeletal—plasma membrane complex. Spectrin is made up of many homologous triple-helical segments joined by nonhelical regions (Speicher and Marchesi, 1984). Reproduced from Malhotra and Shnitka (1991).

and dystrophin (Wagner et al., 1993). It is worth noting that dystrophin contains four potential transmembrane helices in a segment between the 3101 and 3200 residues which could anchor the protein into the membrane (Suwa et al., 1993). Dystrophin and its associated glycoproteins in the sarcolemma interact with the extracellular matrix protein laminin. The absence of dystrophin in patients with Duchenne muscular dystrophy (DMD) results in necrosis of muscle cells. Recent studies of 17 DMD cases of various ages indicate that in addition to the loss of dystrophin, all of the proteins that bind to dystrophin are greatly reduced (up to 90%), based on correlated immunohistochemistry and immunoblotting data (Ohlendieck et al, 1993).

Another well-known instance of cellular abnormality resulting from genetic defects in plasma membrane associated proteins is represented by certain hemolytic anemias in which the red blood cells become mechanically fragile, have abnormal morphology and are prematurely removed from the circulation (see Palek and Sahr, 1992; Peters and Lux, 1993). One such defect is represented by hereditary spherocytosis in which red blood cells are deficient in the peripheral membrane protein spectrin (α, β), a major cytoskeletal protein in these cells. Spectrin is attached to the integral membrane protein band 3 protein through two peripheral membrane proteins, ankyrin and band 4.2. Spectrin is also linked to actin and band 4.1 protein, tropomyosin, and adducin. Band 4.1 protein is also attached to glycophorin which together with band 3 make up nearly one-half of the integral membrane proteins (Figure 3). Band 3 also has binding sites for hemoglobin and glycolytic enzymes (see Darnell et al., 1990). Proteins similar to spectrin, ankyrin, and band 4.1 are present in most nonerythroid cells, where they appear to serve roles similar to those in red blood cells, i.e., in interactions between cytoskeletal constituents and the plasma membrane, for the purpose of regulating cell locomotion and responses to various external stimuli (Luna and Hitt, 1992).

CELL-TO-CELL JUNCTIONS

The plasma membranes of adjacent cells in animal tissues establish specialized cell- to-cell junctions that bind cells to each other and to the extracellular matrix (desmosomes and hemidesmosomes), seal off lumens from surrounding blood spaces as in the intestine, kidney tubules and pancreatic acini (tight junctions or *zonulae occludens*), or allow the passage of ions and small molecules across apposed cells (gap junctions). Only gap junctions are mentioned briefly here, because of their postulated roles in development, differentiation, electrical transmission (electrical synapses), and general cell physiology.

Gap junctions are intercellular channels termed connexons which connect apposed cells (Figures 4-7); each channel is made up of an alignment of two sets of six subunits (one set from each contributing cell). The subunits thus far are known to be made up of a single protein termed connexin. The main types of connexins are liver type (β_I or connexin 32, and β_2 or connexin 26), heart type (α or connexin 43), and lens type (α_3, MP70). The related, but distinct genes which regulate connexins, are expressed in cell-, tissue-, developmental-, or differentiation-stage-specific patterns. While a large number of connexins are now known from different cell types, each cell type generally forms gap junctions from a single type of connexin. For example, in tissues of the mammalian central nervous system and the reproductive system, different cell types form gap junctions from different connexins. Moreover, in these tissues connexins undergo developmental changes (Dermietzel et al., 1989). Also, in the reproductive system, changes in the expression of gap junctions have been correlated with alterations in hormone levels occurring during pregnancy in the rat (Risek et al., 1992).

Figure 4. Electron micrograph showing intercalated discs (gap junctions) in a thin section in rat heart muscle.

The opening and closing of gap junctional channels (cell-to-cell coupling) has been demonstrated to be regulated by changes in voltage, pH, phosphorylation, and Ca^{2+}. The full range of signals and/or factors that are transmitted through gap junction channels remains to be elucidated. cAMP has been shown to pass through the gap junctions formed in cocultures of heart muscle cells and ovarian granuloma cells (Lawrence et al., 1978). Also, based upon studies using diverse approaches including intracellular injections of radioactive or

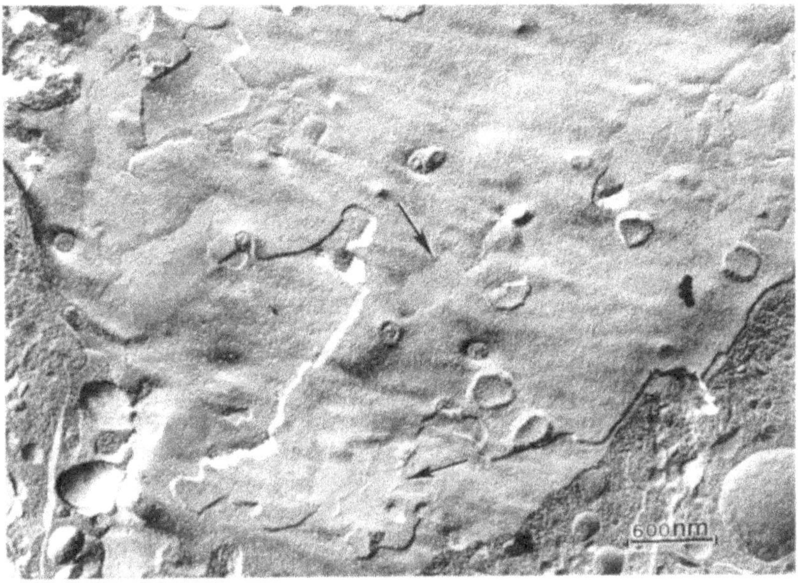

Figure 5. Electron micrograph showing several gap junctions (*arrows*) in a freeze-fracture replica of rat liver.

fluorescent tracers, it is recognized that small molecules and metabolites of up to approximately 1kD pass though gap junctions. Yet the biological significance of the molecular diversity of expressions of gap junctions in various cell types is not yet known. In insect epidermis, ions and lucifer yellow (MW 450) pass freely between cells within the same segment of the insect, whereas ions can move across cells at the segmental boundary, but the dye (lucifer yellow) cannot. Apparently the selectivity of gap junctions at the segmental boundary is different from that within a segment. It indeed would be fascinating to know what factors are responsible for such developmental differences in gap junctions (Warner and Lawrence, 1982).

CONCLUDING COMMENTS

This brief survey stresses the importance of plasma membrane proteins in carrying out numerous cellular functions at the interface between the

Figure 6. Characteristic aggregation of intramembrane particles forming a gap junction in a freeze-fracture preparation similar to that shown in Figure 5 (Sikerwar et al., 1981).

internal and external environments. Future research on regulatory mechanisms of such functions promises to be intellectually exciting and relevant to certain membrane-targeting diseases.

ACKNOWLEDGMENTS

Research grants awarded by NSERC, Canada are gratefully acknowledged.

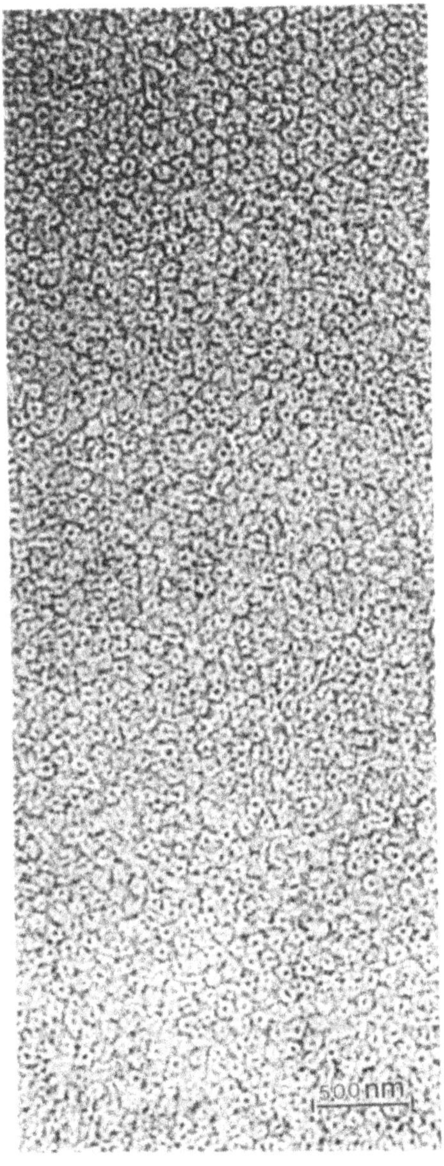

Figure 7. A negatively stained gap junction in a fraction isolated from mouse liver showing packed channels; each channel is filled with electron dense stain seen here as a central dot in a connexon (Sikerwar and Malhotra, 1983).

REFERENCES

Aitken, J. (1992). A family of fusion proteins. Nature 356, 196-197.

Blobel, C.P., Wolfsberg, T.G., Turck, C.W., Myles, D.G., Primakoff, P., & White, J.M. (1992). A potential fusion peptide and an integrin ligand domain in a protein active in sperm-egg fusion. Nature 356, 248-252.

Cowan, S.W., Schirmer, T., Rummel, G., Steiert, M., Ghoshi, R., Pauptit, R.A., Jansonius, J.N., & Rosenbusch, J.P. (1992) Crystal structures explain functional properties of two *E. coli* porins. Nature 358, 727-736.

Darnell, J.E. & Lodish, H., & Baltimore, D. (1990). Molecular Cell Biology, 2nd ed. W.H. Freeman, New York.

Deisenhofer, J. & Michel, H. (1989). The photosynthetic reaction center from the purple bacterium *Rhodopseudomonas viridis*. Science 2, 1463-1474.

Dermietzel, R., Traub, O., Hwang, T.K., Beyer, E., Bennett, M.V., Spray, D.C., & Willecke, K. (1989). Differential expression of three gap junction proteins in developing and mature brain tissues. Proc. Natl. Acad. Sci. USA 86, 10148-10152.

Froehner, S.C. (1991). The. submembrane machinery for nicotinic acetylcholine receptor clustering. J. Cell Biol. 114, 1-7.

Gennis, R.B. (1989). Characterization and structural principles of membrane proteins. In: Biomembranes. Molecular Structure and Function, pp. 85-137. Springer-Verlag, New York.

Gorter, E. & Grendel, R. (1925). On biomolecular layers of lipoid on the chromocytes of the blood. J. Exp. Med. 41, 439-443.

Greunberg, J. & Emans, N. (1993). Annexins in membrane traffic. Trends in Cell Biol. 3, 224-227.

Hynes, R.O. (1992). Integrins: versatility, modulation, and signalling in cell adhesion. (1974). Form and function of the glycocalyx on free cell surfaces. Phil. Trans. R. Soc. Lond. B. 268, 55-66.

Kumar, N.M. & Gilula, N.B. (1992). Molecular biology and genetics of gap junction channels. Seminars in Cell Biol. 3, 3-16.

Lawrence, T.S., Beers, W.H. & Gilula, N.B. (1978). Transmission of hormonal stimulation by cell-to-cell communication. Nature, 272, 501-506.

Lodish, H.F. (1988). Multi-spanning membrane proteins: How accurate are the models? Trends in Biochem. Sci. 13, 332-334.

Luna, E.J. & Hitt, A.L. (1992). Cytoskeleton-plasma membrane interactions. Science 258, 955-964.

Malhotra, S.K. (1983). The Plasma Membrane. John Wiley, New York.

Malhotra, S.K. & Shnitka, T.K. (1991). The cytoskeleton-microtubules and micro-filaments; A biological perspective. In: Fundamentals of Medical Cell Biology (Bittar, E.E., Ed.), Vol. 2. JAI Press, Greenwich, CT.

McCormick, F. (1993). How receptors turn ras on. Nature 363, 15-16.

Mooseker, M.S. (1985). Organization, chemistry, and assembly of the cytoskeletal apparatus of the intestinal brush border. Ann. Rev. Cell Biol. 1, 209-241.

Motulsky, A.G. (1986). The 1985 Nobel Prize in physiology or medicine. Science 231, 126-128.

Nelson, W.J. (1992). Regulation of cell surface polarity from bacteria to mammals. Science 258, 948-955.

Nageli, C. (1855). Quoted in Malhotra (1983).

Nudel, U., Zuk, D., Einat, P., Zeelon, E., Levy, Z., Neuman, S. & Yaffe, D. (1989). Duchenne muscular dystrophy gene product is not identical in muscle and brain. Nature 337, 76-78.

Ohlendieck, K., Matsumura, K., lonasescu, V.V., Towbin, J.A., Bosch, E.P., Weinstein, S.L., Semett, S.W., & Campbell, K.P. (1993). Duchenne musclar dystrophy: Deficiency of dystrophin-associated proteins in the sarcolemma. Neurol. 43, 795-800.

Palek, J. & Sahr, K.E. (1992). Mutations of the red blood cell membrane proteins: From clinical evaluation to detection of the underlying genetic defect. Blood 80, 308-330.

Pazin, M.I. & Williams, L.T. (1992). Triggering signaling cascades by receptor tyrosine kinases. Trends in Biol. Sci. 17, 374-378.

Peters, L.L. & Lux, S.E. (1993). Ankyrins: Structure and function in normal cells and hereditary spherocytes. Seminars in Hematology 30, 85-118.

Phillips, W.D., Kopta, C., Blount, P., Gardner, P.D., Steinbach, J.H. & Merlie, J.P. (1991). ACh receptor-rich membrane domains organized in fibroblasts by recombinant 43-kilodalton protein. Science 251, 568-570.

Rambourg, A. & Leblond, C.P. (1967). Electron microscope observatons on the carbohydrate-rich cell coat present at the surface of cells in the rat. J. Cell Biol. 32, 27-53.

Risek, B., Klier, F.G., & Gilula, N.B. (1992). Multiple gap junction genes are utilized during rat skin and hair development. Development 16, 639-651.

Shariff, A. & Luna, E.J. (1992). Diacylglycerol-stimulated formation of actin nucleation sites at plasma membranes. Science 256, 245-247.

Sikerwar, S.W., Tewari, J.P. & Malhotra, S.K. (1981). Subunit structure of the connexons in hepatocyte gap junctions. Eur. J. Cell Biol. 24, 211-215.

Sikerwar, S.S. & Malhotra, S.K. (1983). A structural characterization of gap junctions isolated from mouse liver. Cell Biol. International Reports 7, 897-903.

Singer, S.J. & Nicolson, G.L. (1972). The fluid mosaic model of the structure of cell membranes. Science 175, 720-731.

Singer, S.J. (1992). The structure and function of membranes—A personal memoir. J. Membrane Biol. 129, 3-12.

Speicher, D.W. & Marchesi, V.T. (1984). Erythrocyte spectrin is comprised of many homologous triple helical segments. Nature 311, 177-180.

Suwa, M., Mitaku, S. & Kuroda, Y. (1993). Theoretical analysis of amino acid sequence of human dystrophin. Biochem. Biophys. Res. Commun. 191, 782-789.

Unwin, N. & Henderson, R. (1984). The structure of proteins in piological membranes. Scientific American 250, 78-86.

Unwin, N. (1986). Is there a common design for cell membrane channels? Nature 323, 12-13.

Unwin, N. (1993). The nicotinic acetylcholine receptor at 9 Å resolution. J. Mol. Biol., 229, 1101-1124.

Wagner, K.R., Cohen, J.B., & Huganir, R.L. (1993). The 87K postsynaptic membrane protein from Torpedo ss a protein-tyrosine kinase substrate holologous to dystrophin. Neuron 10, 511-522.

Warner, A. & Lawrence, P.A. (1982). Permeability of gap junctions at the segmental border in insect epidermis. Cell 28, 243-252.

Warner, A. (1988). The gap junction. J. Cell Sci. 89, 1-7.

Watts, C. & March, M. (1992). Endocytosis: What goes in and how? J. Cell Sci. 103, 1-8.

White, J.M. (1992). Membrane fusion. Science 258, 917-924.

Wolfe, S.L. (1993). Molecular and Cellular Biology. Wadsworth Publishing, Belmouth, CA.

Zachary, I. & Rozengurt, E. (1992). Focal adhesion kinase ($pl25^{FAK}$): A point of convergence in the action of neuropeptides, integrins, and oncogenes. Cell 71, 891-894.

Chapter 2

The Transport of Macromolecules Across the Nuclear Envelope

N. Pokrywka, D. Goldfarb, M. Zillmann, and A. DeSilva

Principles of Medical Biology, Volume 2
Cellular Organelles, pages 19-54

ISBN:1-55938-803-X

INTRODUCTION

The aim of this chapter is to use current notions and hypotheses to illustrate our understanding of nuclear transport. This approach allows the description of instructive principles. Armed with these ideas, students should be able to ponder biomedical problems from a rational basis. For example, we know that the infection of cells by influenza or HIV-1 viruses requires that the viral genomes enter the nucleus to replicate. In this chapter we use HIV-1 nuclear transport as a paradigm to illustrate key principles of nuclear transport, and to demonstrate why understanding an intracellular process is relevant to combating disease. Several excellent reviews about retroviruses, HIV, and lentiviruses appear in Fields' Virology (1985). To aid in conceptualization, it is often helpful for biologists to anthropormorphize cellular events. Thus we can try to empathize both with the host cell and with the renegade virus particle making its desperate run through the cytoplasm to the safety of its nest in the nucleus. To do this, we need to familiarize ourselves with the intracellular environment and how nuclear transport normally proceeds in the absence of a viral attack.

The contents of the nucleus and cytoplasm are determined by their separate functions, and thus, are mostly distinct. The nucleus contains the cell's DNA, and hundreds of different proteins required for structural integrity and myriad processes such as DNA replication, RNA transcription and processing, and ribosome biogenesis. Many of these nuclear activities are thought to occur attached to a solid-state scaffolding rather than in free solution. But the nucleus also contains an aqueous phase, called the nucleoplasm, which contains soluble proteins, ions, and small metabolites such as ATP.

The central problem of nuclear transport is quite simple. All the proteins that are localized in the nucleus are first synthesized in the cytoplasm and must be imported to the nucleus. Likewise, all cytoplasmic RNAs (mainly mRNAs, rRNAs, and tRNAs) are transcribed in the nucleus and, following processing events, are exported across the nuclear envelope. These homeostatic biosynthesis and transport processes are necessary for net cell growth in proliferating cells and for renewing cell constituents that are damaged or intrinsically unstable.

Another class of nuclear transport processes occurs in response to external effectors such as fluctuating nutrients, changing temperature, or endocrine signals. When cells are stimulated externally by growth factors, a signal must be transmitted through the cytoplasm and across the nuclear envelope to the transcriptional apparatus, which must respond to the message by altering the expression of specific genes. This type of event is very common and often involves the transient import from the cytoplasm to nucleus of signal receptor proteins. The nuclear envelope is a site of constant macromolecular trafficking.

The nuclear transport of lentiviruses such as HIV-1, as well as oncoretroviruses, adenoviruses, influenza viruses, and DNA tumor viruses, which also replicate in the nucleus, are pathological events that are distinct from the cell's normal homeostatic and responsive transport processes. Viruses have evolved to cleverly insinuate themselves into the cell's biogenic machinery. Viruses, in fact, are probably the first biotechnical engineers, co-opting cells to overexpress their gene products—and all without venture capital investment! The study of viral nuclear transport can provide unique perspectives to normal transport which in turn opens new avenues of research. The immediacy of human health problems such as the AIDS epidemic can motivate us to pursue some questions more vigorously.

We begin with the structural basis of nuclear transport because in cell biology structure greatly influences our understanding of function. Then, we will discuss what is known about the transport pathway of import substrates beginning from their site of synthesis in the cytoplasm. The orchestrated transfer through the cytoplasm of different kinds of karyophilic (nucleus seeking) macromolecules involves multiple receptor-mediated targeting pathways. These different targeting pathways ultimately converge at the nuclear envelope (NE) and the nuclear pore complex (NPC). The NPC is the transporter which catalyzes the vectorial translocation of macromolecules across the NE. This discussion will provide us with the prerequisite framework to understand cases of regulated nuclear transport that occur during development and the cell cycle, in response to environmental changes, and during diseases such as HIV-1 infection.

STRUCTURAL BIOLOGY OF NUCLEAR TRANSPORT

The cytoplasm of most eukaryotic cells is enclosed by the plasma membrane and filled with organelles consisting of distinct membrane-bound compartments. The nucleus itself is surrounded by two mem-

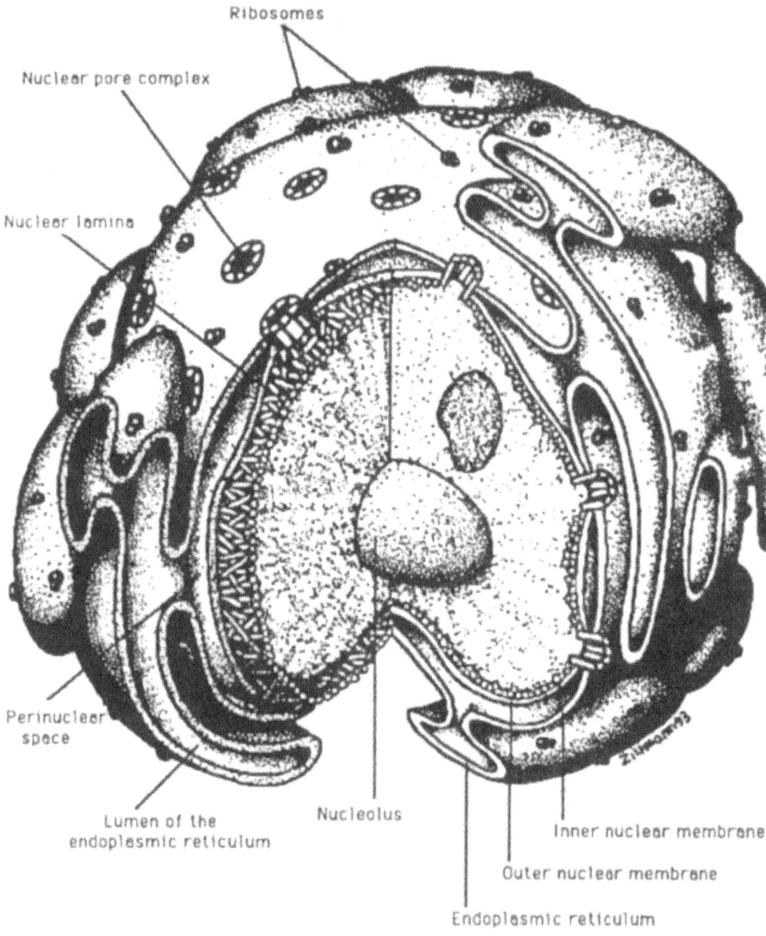

Figure 1. A three-dimensional drawing of the nucleus and associated membrane compartments. A section of the nuclear envelope has been removed to reveal the spatial relationships of the different membranes and compartments.

branes collectively known as the nuclear envelope (NE) (Figure1). The NE prevents the random mixing of the nucleoplasm and the cytoplasm while specialized structures known as nuclear pore complexes (NPCs) occur at sites where the inner and outer nuclear membranes meet and permit the selective transport of molecules between the compartments.

Like prokaryotic and eukaryotic cells, infectious HIV-1 particles are also surrounded by a membrane. Viruses do not, however, have the equipment to replicate themselves independent of cells. They must infect cells in order to proliferate. Upon infection of a cell the HIV-1 membrane fuses with the cell's plasma membrane and the viral core is released into the cytoplasm. In the cytoplasm, after the HIV-1 RNA genome is reverse transcribed into double-stranded DNA, the particle is ready to cross the nuclear envelope. Once in the nucleus the viral genome integrates into the host cell DNA. HIV-1 may employ either of two strategies to get across the nuclear envelope. Before describing these strategies, both of which are used by normal cellular components, we first need to examine the structure and dynamics of the nuclear envelope during interphase and mitosis.

The NE consists of an inner and outer membrane (Figure 1). The inner bilayer is probably in contact with chromosomes and is supported from the inside by a protein meshwork called the nuclear lamina. The outer bilayer is continuous with the membrane of the endoplasmic reticulum (ER). The space between the inner and outer membranes is called the perinuclear space and is equivalent to the lumen of the ER. Due to their topological continuity, the perinuclear space and the ER lumen appear to be similar in composition. Newly synthesized secreted proteins can be found in the perinuclear space as well as the ER lumen. The outer nuclear membrane and the membrane of the ER are similar in their protein and lipid composition and the presence of ribosomes on both membranes indicates that they are sites of membrane and secretory protein synthesis. Of course, the ER lacks the NPCs which are characteristic of the NE. That the inner and outer nuclear membranes are distinct is reflected by the fact that their protein and lipid compositions are different. NPCs are formed at sites where the inner and outer membranes meet. A typical human cell nucleus is penetrated by 2,000-4,000 pore complexes. While this may not seem like many, a yeast nucleus has fewer than 200 NPCs.

Based on its position between the inner and outer nuclear membranes, the NPC must serve as a barrier that prevents the mixing of the two membranes' components. However, it is likely that integral membrane proteins of the inner membrane have to slip by the periphery of the NPC to get from their site of synthesis on the ER-outer NE membranes to the inner membrane. Thus, the NPC may be involved in controlling both the movement of macromolecules between the nucleoplasm and cytoplasm and the movement of molecules between the inner and outer membranes.

The evolutionary origin of the two NE membranes and their relationship to the ER is suggested by structures that are associated with the chromosomes of certain bacteria. In these prokaryotes, the

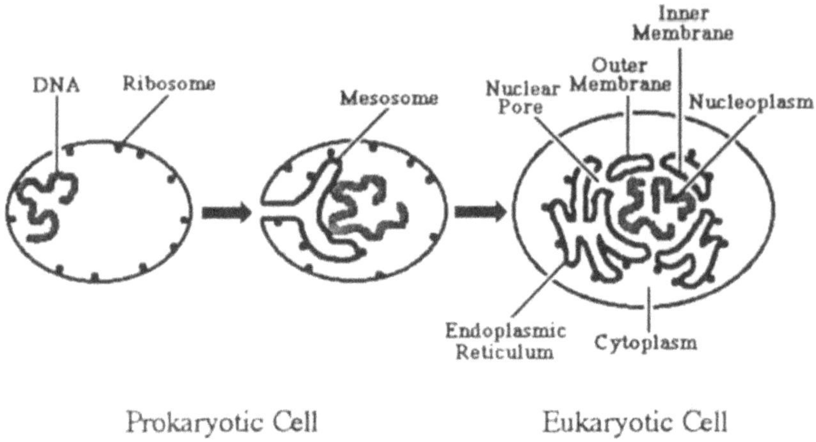

Figure 2. A hypothetical model for the evolutionary origin of the nuclear envelope. In some bacteria DNA is associated with invaginations of the plasma membrane known as mesosomes. These structures may have separated from the plasma membrane to form the nuclear envelope and associated endoplasmic reticulum of eukaryotes. (Adapted from Molecular Biology of the Cell, Bruce Alberts et al., 2nd ed., Garland Publishing, Inc., New York.)

chromosome is associated with specialized invaginations of the cell membrane known as mesosomes. The eukaryotic nuclear compartment may have arisen from gross extensions of these invaginations enclosing the DNA to form the nuclear envelope and the associated ER (Figure 2). This model is attractive because it explains the existence of the two membranes as well as the continuity between the outer nuclear membrane and the ER. There is little doubt that the eukaryotic protein secretion machinery located in the ER is descended from bacterial secretory systems located in the cell membrane, since the two apparatus contain homologous proteins. Mysteriously, we have no clues as to the evolutionary origin of the NPC.

The Nuclear Lamina is Associated with the Inside of the Inner Nuclear Membrane

On the nucleoplasmic side of the NE there is a supporting fibrous meshwork named the nuclear lamina (Forbes, 1992). The lamina consists of one or more lamin proteins. In mammals, each of the three lamins A, B, and C have been cloned and extensively studied. In addition to the three major mammalian lamins, homologous proteins have been found in a variety of other eukaryotic organisms, although not yet in the yeasts. All three mammalian lamins are peripheral membrane proteins. Thus, they are synthesized on free ribosomes in the cytoplasm, and transported as soluble proteins into the nucleus where they are woven together to form the lamina. Lamin B is bound tightly to the inner nuclear membrane via the lamin B receptor, an integral inner membrane protein.

The DNA sequence of the lamins has revealed that they are related to intermediate filament proteins such as keratin and vimentin. Intermediate filaments typically confer structural support to the cell. This indicates that one of the functions of the nuclear lamina is to help maintain the structural integrity of the NE, much like steel scaffolding supports the weight of a large building. This idea is supported by the examination of purified nuclei, which retain their rigid spherical shape even after removal of their lipid membrane by detergent treatment.

The influence of the spatial organization of the interior of the nucleus on gene expression is a topic cell biologists have recently begun to investigate. The observation that the chromatin which is associated with the NE is predominantly in the form of heterochromatin (which, in contrast to euchromatin, is thought to be actively transcribed) hints at a possible relationship between the attachment of genes to the NE and their expression. The NE may also play a role in organizing the chromosomes within the nucleus. The differential expression of lamins in various cell types and during development may be one way of altering both nuclear architecture and gene expression.

The Nuclear Envelope During Mitosis

A striking reorganization occurs within higher eukaryotic cells upon the onset of mitosis (Laskey and Leno, 1990). In addition to the depolymerization of cytoplasmic microtubules and the appearance of the spindle apparatus, the NE experiences its own striking metamorphosis. During interphase, the surface area of the NE and the number of NPCs is doubled in preparation for cell division. Then, in higher eukaryotes, the nuclear membranes disintegrate into hundreds of small vesicles and the nuclear lamina disassembles into lamin subunits. Lamins A and C are rendered freely soluble in the cytoplasm, while lamin B remains associated with inner membrane vesicles, presumably by maintaining its grip on the lamin B receptor.

The catastrophic collapse of the NE results in the wholesale mixing of cytoplasmic and nuclear components. This randomization process has at least one function. It assures that cellular components are equally divided between the daughter cells based on volume. At telophase the reassembly of the two daughter nuclei begins. Initially, lamins A and C associate with the condensed mitotic chromosomes. About the same time, membrane vesicles that contain lamin B bind to the lamin A- and C-coated chromatin. Other vesicles fuse with the lamin B-containing vesicles to form a double envelope around the chromosomes. The nuclear envelope initially forms tightly around the chromosomes and then later expands outward as the chromosomes decondense. This close association of the membrane vesicles with the condensed chromosomes helps to prevent the trapping of cytoplasmic components inside the newly formed nucleus. But this mechanism also excludes all the nuclear components floating around in the larger cytoplasm. These components must reenter the nucleus during the expansion phase of postmitotic NE biogenesis. The mechanism of nuclear import following mitosis is probably the same as the mechanism in operation during interphase.

The assembly and disassembly of the nuclear envelope, together with chromosome condensation, can be studied in *vitro* using cell extracts and purified DNA (Laskey and Leno, 1990). These reconstitution assays have allowed the step-by-step dissection of the complicated restructuring of nuclear components. One of the conclusions from these studies is that the formation of small nuclear membrane vesicles, the disintegration of the nuclear lamina into individual lamins, and the condensation of chromosomes are independent reactions that can be studied separately. The phosphorylation and dephosphorylation of nuclear proteins initiates and then coordinates nuclear assembly and disassembly. For example, the hyperphosphorylation of the lamins leads directly to their solubilization. Lamins will reassemble into a meshwork only when these excess phosphate groups are removed. Some lower eukaryotes, yeasts and dinoflagellates, for example, go through the cell cycle with their nuclear envelopes mostly intact. Here, instead of reforming independently in the two daughter cells, the nuclei of these primitive cells appear to pinch in half to create two daughter nuclei. This difference may one day be exploited to preferentially attack the cell cycle of infectious yeasts such as *Candida* species.

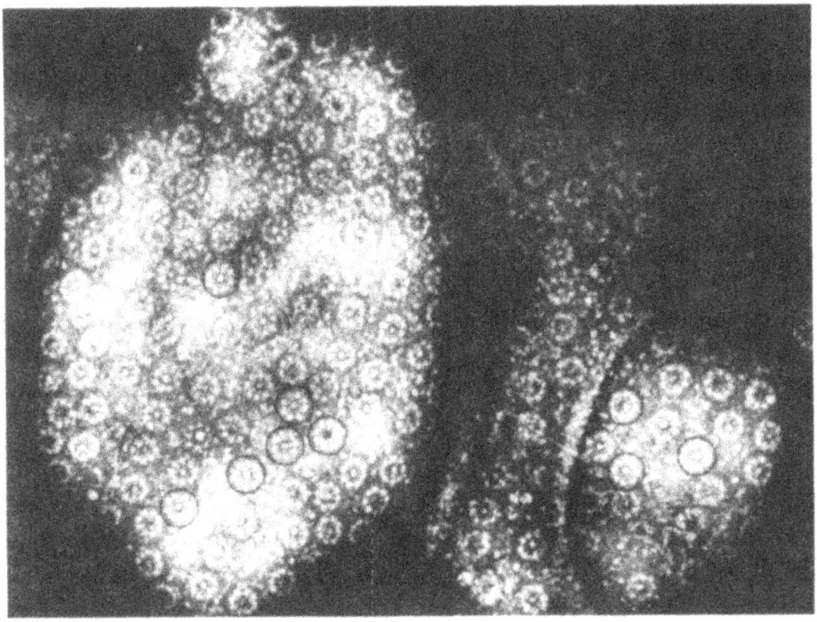

Figure 3. Electron micrograph of the amphibian oocyte nuclear envelope. This micrograph indicates the density of nuclear pore complexes on the oocyte nuclear envelope. The eightfold symmetry of the pore complexes is apparent. Figure provided by R. Milligan.

It is during the reassembly of the nucleus at telophase that oncoretroviruses gain access to the nucleus. The uncoated virions (containing double-stranded DNA) are apparently mistaken as nuclear components, perhaps chromosomes, and become packaged within the newly formed daughter nuclei. This is an essential mechanism for most retroviruses because they lack the tools to cross the intact NE of interphase cells. In contrast, the lentiviruses can enter the nucleus both during mitosis and interphase. This latter capacity is critical for the HIV-1 infection of nonproliferating cells such as circulating macrophages and central nervous system cells, most of which are arrested in interphase. Clearly, the study of nuclear transport in healthy cells has laid both the conceptual and experimental foundation for our investigations on the mechanism of HIV-1 nuclear transport.

The Nuclear Pore Complex

The NPC is the centerpiece of nuclear transport. It is a huge structure with a mass of 125 million daltons, which makes it about 34 times the size of a ribosome. The NPC is large enough to appear distinctly in electron micrographs of the NE surface (Figure 3). The major elements of the amphibian oocyte NPC have been resolved by the computer enhancement of imperfect images taken with the electron microscope (Hinshaw et al., 1992; Akey and Radermacher, 1993). These processed images portray the NPC as a structure with eightfold symmetry in the plane of the NE and rough twofold symmetry across the envelope (Figure 4). The oocyte NPC has a diameter of about 150 nm and a thickness of 70 nm. Various laboratories agree that the NPC is comprised of cytoplasmic and nucleoplasmic ring structures connected by massive spokes. The ring and spoke assembly leaves enough space in the middle for the transport of macromolecules as large as ribosomal subunits and virions. The organization of the middle domain of the NPC is poorly understood. This is unfortunate because it is here where the elusive transport mechanism resides. Attached to the nucleoplasmic side of the NPC is a large basketlike structure of unknown function. Equally mysterious is the appearance in some electron micrographs of filaments which protrude from the cytoplasmic face of the NPC. The structural biologists have provided us with images that portray a large and unique transport apparatus of great sophistication and complexity.

Proteins of the Nuclear Pore Complex

One important approach to understanding how the NPC functions is to isolate and characterize the individual proteins of the NPC. Proteins of the NPC are called nucleoporins. Based on its mass, it has been estimated that over 100 different nucleoporins make up the NPC. An approach that has been successfully used to identify and clone a few of the NPC proteins begins by obtaining relatively pure NPC preparations and raising monoclonal antibodies to these crude fractions (Forbes, 1992). Antibodies directed against NPC proteins were identified by selecting those that stained the NPC by indirect immunofluorescence. These antibodies were then used to purify individual NPC proteins. With the characterized antibodies and purified nucleoporins in hand, "reverse genetic" techniques were used to clone the nucleoporin genes.

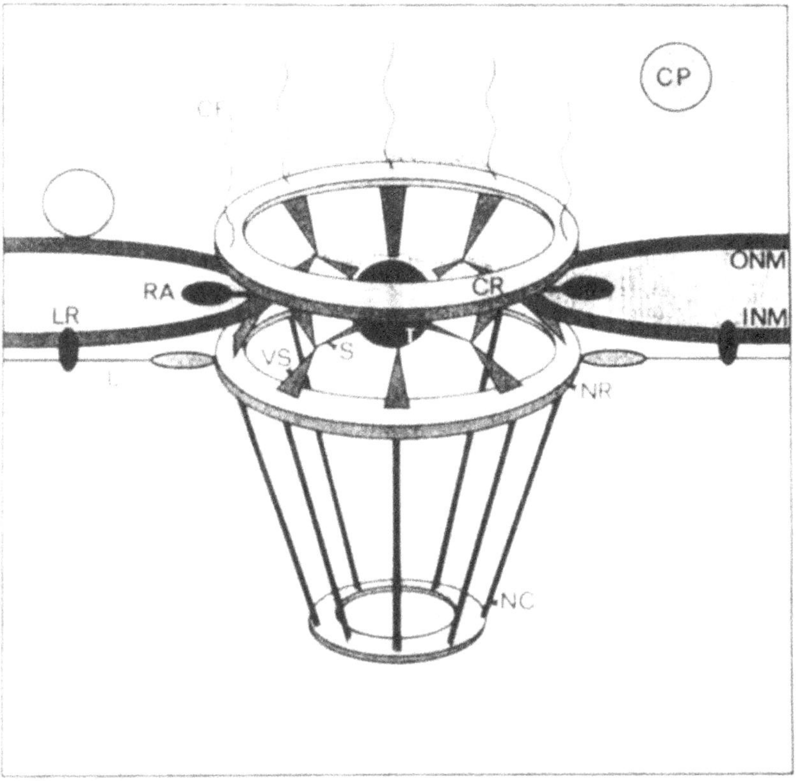

Figure 4 Structure of the nuclear pore complex. The components of the NPC include the nuclear and cytoplasmic rings (NR and CR), spokes (S), vertical supports (VS), radial arms (RA), cytoplasmic filaments (CF), and nuclear cage or basket (NC). The NPC is attached to the nuclear lamina (L), which is attached to the inner nuclear membrane (INM) via lamin receptors (LR). The outer nuclear membrane (ONM) is sometimes decorated with cytoplasmic particles (CP). Figure provided by C. Akey.

These genes are now being dissected in order to ascertain the molecular mechanism of nucleoporin function.

Some of these monoclonal antibodies recognize multiple proteins indicating that the epitope recognized may be shared by a family of related nucleoporins. Some of the vertebrate nucleoporins identified to date are post-translationally modified by the addition of serine and threonine-linked N-acetylglucosamine (Forbes, 1992). These glycosylations are catalyzed by cytosolic enzymes and thus are distinct from those which occur within the lumen of the ER and Golgi apparatus. The lectin wheat germ agglutinin (WGA), which binds to N-acetylglucosamine residues, also blocks nuclear transport. Anti-nucleoporin monoclonal antibodies, which bind to the same polypeptides as WGA does, also inhibit nuclear transport. Inhibition of transport by WGA and antibodies indicates that the nucleoporins play a functional role in transport. Nucleoporins that are responsible for anchoring the NPC into the membranes of the NE have also been identified and, as expected, these are integral membrane proteins that protrude into the perinuclear space.

One of the more prominent nucleoporins identified was a 62 kD polypeptide (p62). To date, p62 genes have been cloned from rat, frog, and humans, and a distant relative has been identified in yeast. The p62 protein consists of three domains: (1) an N terminal region containing 12 repeats of a 7 amino acid sequence; (2) a central domain rich in Ser, Thr, Pro, and Ala residues, and (3) a unique C terminal domain. The 7 amino acid repeats are also found in other nucleoporins, although their function is unknown. Recent studies with rat NPC preparations have revealed that p62 exists in a tight 550-600 kD complex with nucleoporins of 58 and 54 kD. Identifying the interactions between various nucleoporins is important for establishing the gross molecular structure of the NPC. The identification of nucleoporins in yeast is especially exciting because the powerful genetic manipulations possible in yeast can be used to analyze the structure and function of these proteins. As more NPC proteins are characterized, the goal of understanding how the NPC functions as a macromolecular transporter will become a reality.

The Nuclear Pore Complex is the Site of Nucleocytoplasmic Exchange

A classic experiment in nuclear transport used electron microscopy to directly demonstrate that proteins are transported through the NPC (Feldherr et al., 1984). Colloidal gold particles were coated with nuclear proteins (karyophilic gold particles) and microinjected into the cytoplasm of frog oocytes. The subcellular distribution of the electron-dense gold particles was then examined using electron microscopy. Karyophilic gold particles accumulated in the nucleus and, significantly, were also observed in transit through the NPCs. The import of rigid gold particles also proved that macromolecules do not need to be unfolded or significantly deformed during transport. This characteristic allows for the transport of macromolecular assemblages such as ribosomal subunits, RNA-protein complexes, and, as it happens, viral particles.

Experiments of this type have taught us other things about the functional capabilities of the NPC. The largest macromolecule capable of being carried across the NPC was estimated by microinjecting sized karyophilic gold particles. Interestingly, the capacity of the cell to import larger karyophiles is sensitive to the physiological state of the cell (Feldherr and Akin, 1991). In healthy proliferating cells, the maximum diameter appears to be about 26 nm. Twenty six nm is approximately the diameter of the largest physiological export substrate, the 60S ribosomal subunit. This point highlights a salient feature of nuclear transport: the NPC is the site of both nuclear import and export. It is not yet understood how the NPC achieves the bidirectional transport of proteins and RNAs (see below).

Other microinjection studies discovered a surprising fact about the transport of relatively small macromolecules (Paine et al., 1975). Molecules smaller than about 40 kD can freely diffuse into the nucleus through 9 nm diameter aqueous pores in the NPC. Later it became clear that these sieving properties are mostly irrelevant because most macromolecules of mass less than 40 kD exist in the cell as protein-protein complexes that are larger than this 9 nm size cut-off (Breeuwer and Goldfarb, 1990).

NUCLEAR LOCALIZATION SIGNALS

One of the most significant principles of nuclear transport came with the discovery in the 1980s that amino acid sequences called nuclear localization signals (NLS) direct the import of nuclear proteins. The first indication of a NLS domain came from microinjection studies in frog oocytes (Dingwall et al., 1982). Nucleoplasmin, which is an abundant nuclear protein of 150 kD, can be proteolytically cleaved into N-terminal head and C-terminal tail pieces. When microinjected into the cytoplasm of oocytes, the tail fragments moved into the nucleus while the head fragments remained in the cytoplasm. It was concluded that the tail piece is the domain that directs the movement of the intact nucleoplasmin molecule into the nucleus.

The first discrete NLS was discovered within the large T-antigen protein of the monkey virus SV40 (Garcia-Bustos et al., 1991; Forbes,

1992). SV40 is a virus which replicates and is transcribed in the nucleus. A large amount of T-antigen is expressed in infected cells and, since it functions to control viral transcription and replication, this newly synthesized protein is imported to the nucleus much like other nuclear proteins. When mutations were made in a basic region of the large T-antigen protein, at amino acid 128 (a lysine residue), the mutant T-antigen localized to the cytoplasm instead of the nucleus. A sequence of seven amino acids surrounding lysine-128 were subsequently shown to be both necessary and sufficient to promote nuclear localization. This short sequence is, in fact, the NLS of T-antigen.

NLSs in many nuclear proteins have been identified by deletion analysis, which results in the cytoplasmic accumulation of the nuclear protein, and by insertion of NLS sequences, which cause the nuclear localization of a cytoplasmic protein. The total autonomy of NLS sequences was demonstrated by showing that short synthetic peptides containing NLS sequences could direct the import of virtually any macromolecule to which they were attached.

An examination of most NLS sequences does not reveal an obvious linear consensus sequence (Table 1). Some proteins contain more than one NLS sequence which may interact to provide for more efficient import. Most NLS sequences contain basic amino acids, arginines and lysines, and many have one or more prolines or glycines at their borders. Prolines and glycines are important in determining the structure of polypeptides and may act here to assure the accessibility of the NLS to NLS receptors. Most NLS sequences are at least vaguely related either to the archetypal SV40 large T-antigen NLS (PKKKRKV) or the nucleoplasmin NLS that is comprised of two domains of basic amino acids separated by a spacer of ten amino acids (KRPAATKKAGQAKKKK). Because the double basic domain nucleoplasminlike NLS is more common, some believe that the T-antigen NLS may be an extreme single-ended variant of the bipartite nucleoplasmin NLS. Both the single basic domain T-antigen and double basic domain nucleoplasmin motifs should be considered together because they appear to be functionally indistinguishable, although some are "stronger" than others. The basic domain NLSs are evolutionarily conserved, as shown by the ability of the monkey T-antigen NLS to target proteins to the nuclei of yeast and *Tetrahymena*, a ciliated protozoan. NLS sequences can occur anywhere within the protein, provided the NLS is accessible and not buried within the folded polypeptide chain. An outstanding question in the area of viral nuclear

Table 1. Nuclear localization signal sequences from a variety of organisms. Basic amino acids are in bold type, proline and glycines are underlined. Some proteins listed have two or three NLSs.

Organism	*Protein*	*NLS Sequence*
Human	hsp70	FKRKHKKDISQNKRAVRR
	lamin A	TKKRKLE
	PDGF A	PRESGKKRKRKRLKPT
	c-myc	PAAKRVKLD
	N-myc	PPQKKIKS
	c-myb	PLLKKIKQ
Monkey	PDGF-B	RVTIRTVRVRRPPKGKHRK
Mouse	c-abl	KKKKK
Rat	glucocorticoid receptor	RKTKKKIK
Rabbit	progesterone receptor	RKFKKFNK
Chicken	Ets1	GKRKNKPK
	v-rel	KAKRQR
Frog	N1	VRKKRKT
	lamin L1	VRTTKGKRKRIDV
	nucleoplasmin	KRPAATKKAGQAKKKKLD
Corn	R protein	GDRRAAPARP
		MSERKRREKL
		MISESLRKAIGKR
Yeast	ribosomal L3	PRKR
	ribosomal L29	KTRKHRG
		KHRKHPG
	histone H2B	GKKRSKA
	MATa2	KIPIK
Adenovirus	pTP	RLPVRRRRRRVP
	DBP	PPKKR
	E1a	KRPRP
BPV	E1	LKRK
		KRRLF
Epstein-Barr Virus	nuclear antigen 3A	RDRRRNPASR
Influenza Virus	nucleoprotein	AAFEDLRVRS
	NS1	DRLRR
HIV-1	Tat	GRKKR
	Rev	RRNRRRRW
HTLV-1	Rex	PKTRRRP
SV40	large T antigen	PKKKRKV
	VP1	APTKRK
	VP2	PNKKKRK

Source: Adapted from Garcia-Bustos et al., 1991.

transport is the identity and location of NLS sequences that are responsible for the import of viral genomes.

THE PATHWAY OF NUCLEAR IMPORT

Nuclear import can be divided into two biochemically and temporally distinct stages (Garcia-Bustos et al., 1991; Forbes, 1992). First, the karyophile is targeted from its site of synthesis through the cytoplasm to the NPC. While this targeting phase may be diffusion-driven, a number of protein factors, including NLS receptors, are involved. Second, the karyophile is transported through the NPC into the nucleus. The nuclear targeting of most proteins has been shown to proceed by a series of cytosolic ATP-independent binding reactions, followed by the ATP-dependent accumulation of karyophile in the nucleus. Import is thought to be initiated by the formation of a targeting complex between a cytoplasmic NLS-receptor and a newly synthesized NLS-containing protein. By the time the protein has reached the NPC a number of cytosolic factors will have participated. While few of these factors have been characterized, their existence indicates that the cell has made a large commitment to the control of nuclear import (see below). But while the transport apparatus appears to be quite complex, it is important to remember that all a protein needs to link up with this apparatus is a simple NLS sequence.

The nuclear transport of proteins, at least in higher eukaryotes, occurs after their translation and folding in the cytoplasm. Following their import, nuclear proteins retain their NLS sequences. In contrast, secretion signals and mitochondrial targeting signal sequences are proteolytically removed. Why do nuclear proteins retain a sequence that is only used during transport? As described above, in higher eukaryotes the nuclear envelope breaks down every time the cell passes through mitosis. During mitosis, then, many nuclear proteins become distributed throughout the cytoplasm. Because these proteins retain their NLSs they are able to reenter the newly formed nuclei. This is an interesting speculation but it does not explain why yeast NLSs are retained even though yeast nuclear envelopes do not appear to break down during mitosis.

Certain HIV-1 proteins contain amino acid sequences that look like basic domain NLSs (Table 1). One or more of these NLSs may be responsible for the import of the large HIV-1 core particle across the NE of interphase cells. In fact, the deletion of an NLS from the HIV-

Table 2. Virus families which spend part of their life cycle in the host nucleus.

Family	*Examples*	*Genome*
Orthomyxoviridae	influenza virus	ss RNA
Retroviridae	retroviruses	*ss RNA
	oncoviruses	
	HIV	
Hepadnaviridae	hepatitis B-like viruses	ss/ds DNA
Parvoriridae	parvoviruses	ss DNA
	papilloma viruses	
Papovaviridae	polvoma viruses	ds DNA
Adempvorodae	adempvorises	ds DNA
Herpesviridae	herpes simplex virus 1 & 2	ds DNA
Iridoviridae	insect and frog viruses	ds DNA

Note: *Genome is reverse transcribed into ds DNA which then enters the nucleus

1 gag-matrix protein prevents one strain of HIV-1 from entering the nucleus of cells arrested in interphase (Bukrinski et al., 1993) This defective virus can still enter the nuclei of dividing cells by using the postmitotic enclosure mechanism described above. It is likely that nonlenti retroviruses are unable to infect nonproliferating cells because they lack the requisite NLS sequences. The acquisition of a serviceable NLS by lentiviruses appears to have led to deadly diseases of nonproliferating cells of mammalian central nervous and immune systems. These include the maedi-visna disease discovered and studied in Icelandic sheep and, of course, AIDS in humans (Table 2).

FACTORS INVOLVED IN NUCLEAR IMPORT

Three approaches are being used to identify cellular components of the nuclear transport apparatus. First, NLS receptors have been sought by searching for proteins that bind tightly to synthetic NLS peptides (Yamasaki and Lanford, 1992). Second, *in vitro* transport assays are being used to fractionate cell free extracts and reconstitute nuclear transport from purified components (Adam and Gerace, 1991; Moore and Blobel, 1992). And third, yeast genetics are being used to identify important genes by the reverse genetic techniques alluded to above and by selecting for mutant strains that exhibit defective phenotypes in nuclear transport (Osborne and Silver, 1993).

The development of *in vitro* nuclear transport assays that are dependent upon the addition of cell-free extracts for import has allowed investigators to begin to isolate proteins involved in nuclear transport, such as putative receptors, through fractionation and reconstitution. In one of these *in vitro* assays, the plasma membrane of tissue culture cells can be selectively permeabilized by a cholesterol-solubilizing detergent, digitonin. Cholesterol is a major lipid component of the plasma membrane. Digitonin treatment destroys the integrity of the plasma membrane and releases many cytoplasmic components that are required for nuclear transport. These components may be added back in the form of cell-free extracts to restore transport function. Moore and Blobel (1992) used this approach to discover that separate fractions are required for the targeting and translocation steps of import. Fraction A contains an activity that is required for the targeting of karyophile to the nuclear membrane and is inactivated by treatment with N-ethylmaleimide. Fraction B contains a factor or factors that mediate the subsequent ATP-dependent translocation of pre-targeted karyophiles into the nucleus. Fraction A and B each contain scores of different proteins. The task ahead is to use biochemical fractionation techniques and the *in vitro* transport assay to determine which of these proteins are responsible for the separate activities.

Cell biologists are excited about the possibility of employing these *in vitro* assays to the study of virus nuclear transport. If, for example, there exists a drug that specifically inhibits the nuclear transport of HIV-1, then the screening of candidate drugs will be simplified by a convenient *in vitro* import assay. Precedent for this notion comes to us from the case of amantadine, an antiviral drug. Amantadine was shown to block a step along the pathway of influenza virus entry into cells, thereby preventing the import of incoming virus genomes (Martin and Helenius, 1991).

The energetics of translocation is an important issue that remains incompletely resolved. Because translocation could theoretically occur either by active transport, facilitated diffusion, or in the case of small (less than about 40 kD) proteins, by simple diffusion, it is important to experimentally distinguish between these exclusive mechanisms. Several criteria have been used to experimentally define the route of import. Active transport should require energy (usually in the form of ATP), pump substrate against a concentration gradient, and be dependent on a limited number of substrate receptor binding sites. Facilitated diffusion should not require energy, should not operate against a concentration gradient, but should still depend on a limited number of import receptors. Both of these mechanisms can be blocked by NPC binding agents such as wheat germ agglutinin or antinucleoporin antibodies. In contrast, simple diffusion does not require energy, can not pump against a concentration gradient, and is not receptor-mediated. As the diffusion channel of the nuclear pore is only 9 nm in diameter, this last route of entry is available only to macromolecules with an effective diameter below 9 nm. Although the diffusive nuclear transport of selected macromolecules is actively researched, it is a poorly understood process which we will ignore here. Where it has been investigated, the translocation of macromolecules into the nucleus requires ATP and fulfills the other criteria by which we judge a process to occur by active transport.

A difficult problem is determining the exact role of ATP in nuclear transport. It is clear that ATP is required in the cell for proteins to enter the nucleus, but it is unclear at which step in the translocation pathway ATP is used. This question is also one of the outstanding problems in other ATP-dependent membrane transport processes, especially in the area of protein transport across the ER and mitochondrial membranes.

How is the energy of ATP hydrolysis converted into vectorial nuclear transport? There are a surprising number of alternative possibilities. Often in the cell ATP hydrolysis is used to induce a conformational change in a protein which results in a significantly increased or decreased affinity of that protein for a ligand. One example is the ATP-associated heat shock protein hsp70, which binds to unfolded amino acid sequences of newly translating proteins. Upon ATP-hydrolysis, hsp70 changes conformation and the hsp70-protein complex dissociates. Hsp70 has been shown to play a role in nuclear transport (Goldfarb, 1992) and, therefore, its ATPase activity may somehow be essential for the transport process. In light of these speculations, it is important to emphasize that the role of ATP in nuclear transport is unknown.

Multiple Nuclear Targeting Pathways

Although much of the traffic across the nuclear envelope consists of the import of newly synthesized nuclear proteins and the export of recently transcribed RNAs, there are other more complex pathways (Goldfarb and Michaud, 1991). For example, ribosomal proteins are first imported into the nucleus, assembled into ribosomal subunits in the nucleolus, and then exported back to the cytoplasm. Similarly, snRNAs are transcribed in the nucleus, exported to the cytoplasm, assembled in snRNPs, and imported back into the nucleus. Some proteins are continuously transported back and forth through the NPC in a process called shuttling. A difficult logistical problem in nuclear transport is the coordination of import and export pathways. Before we discuss bi-directional nuclear transport, it is useful to learn how multiple import pathways are organized.

It is now clear that the nuclear import of most, if not all, newly translated nuclear proteins is mediated by basic domain NLS sequences. Because these proteins contain homologous NLSs they are recognized by the same NLS receptors and targeted to the NPC by a common pathway. Under conditions of normal homeostatic biogenesis, the cytoplasm contains enough NLS receptor to manage the load of newly synthesized nuclear proteins. If too many nuclear proteins were synthesized in the cytoplasm, then the NLS receptors will become saturated and import substrates will back up in the cytoplasm (much like cars during rush hour). While nuclear traffic jams are not known to occur naturally, we can impose this condition on the cell by microinjecting high concentrations of nuclear proteins into the cytoplasm of cells. This results in a phenomenon called saturation kinetics. Saturation kinetics is a group of techniques used in the analysis of transport pathways and occurs only when transport is mediated by a limiting cellular factor such as a receptor (Figure 5). Passive transport, which is simply the diffusion of molecules through holes in membranes, does not display saturation kinetics because there are no transport factors to saturate. When nuclear transport is saturated by the injection of a single karyophilic species, the import of most nuclear proteins is inhibited. This result argues that most proteins compete for the same NLS-receptor during the targeting phase of nuclear transport. There are, however, exceptions to the single pathway rule.

The story of multiple targeting pathways begins with the import of the uridine-rich small nuclear ribonucleoprotein particles (U snRNPs, pronounced "snurps") (Goldfarb and Michaud, 1991). U snRNPs are assembled in the cytoplasm and imported to the nucleus where they function to mediate the splicing of intron-containing RNAs. U snRNPs are interesting because they appear to use a different targeting pathway than proteins that contain basic domain NLSs. The transport pathway of U1 snRNP is typical of many U snRNPs (Figure 6). U1 snRNA is transcribed by RNA polymerase II, which is also responsible for the synthesis of messenger RNAs. RNA polymerase II transcripts contain a 7-monomethylguanosine cap (7-mGpppG) which is thought to protect capped RNAs from degradative RNAses. Following its export into the cytoplasm, the U1 snRNA is bound by a set of proteins called the Sm

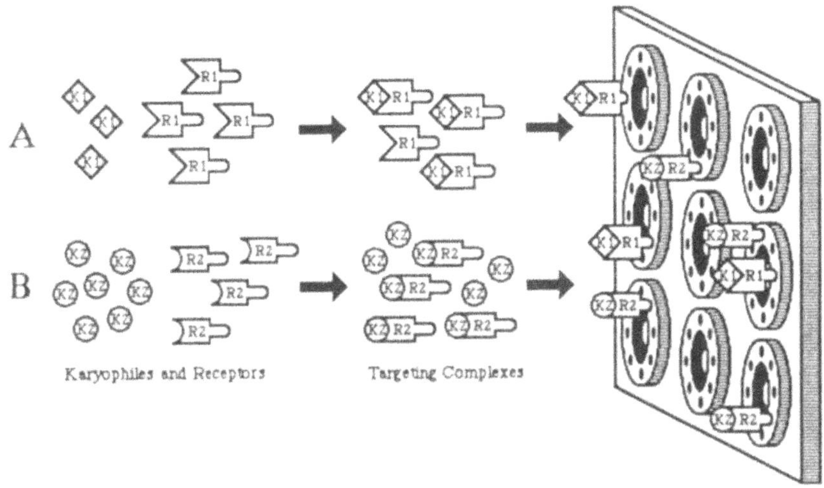

Figure 5. Saturation of import and the basis for kinetically distinct classes of nuclear localization sequences. **A** shows the import of sub-saturating amounts of karyophile 1 (K1), leaving some of its targeting receptor (R1) without ligand. **B** depicts the import of saturating quantities of karyophile 2 (K2) such that no unliganded targeting receptor (R2) remains. Since the receptors are limiting with respect to the availability of pores, all liganded receptors can dock at pores and release their burden into the nucleus. This means that although the import of K2 is proceeding at its maximal rate, there is no effect on the rate of transport of K1.

antigens to form an RNP particle. Finally, two additional methyl groups are added to the 7-methylguanosine cap, producing a 2,2,7-trimethylguanosine cap (TMG), and a few nucleotides at the 3' end of the snRNA are removed. Only the fully assembled and hypermethylated U1 snRNP is recognized by the import apparatus.

Kinetic studies have shown that the import of U1 snRNP proceeds by a unique targeting pathway. Thus, the import of U1 snRNP is not affected by saturating concentrations of proteins containing basic domain NLSs. Conversely, saturating concentrations of U1 snRNP have no effect on the import of basic domain-containing karyophiles. Apparently U1 snRNP contains a distinct NLS that is recognized by its own cognate NLS receptor. What is the U1 snRNP NLS? Results to date indicate that the U1 snRNP NLS is comprised of both the TMG cap and some part of the Sm antigens. In fact, U1 snRNP import can be

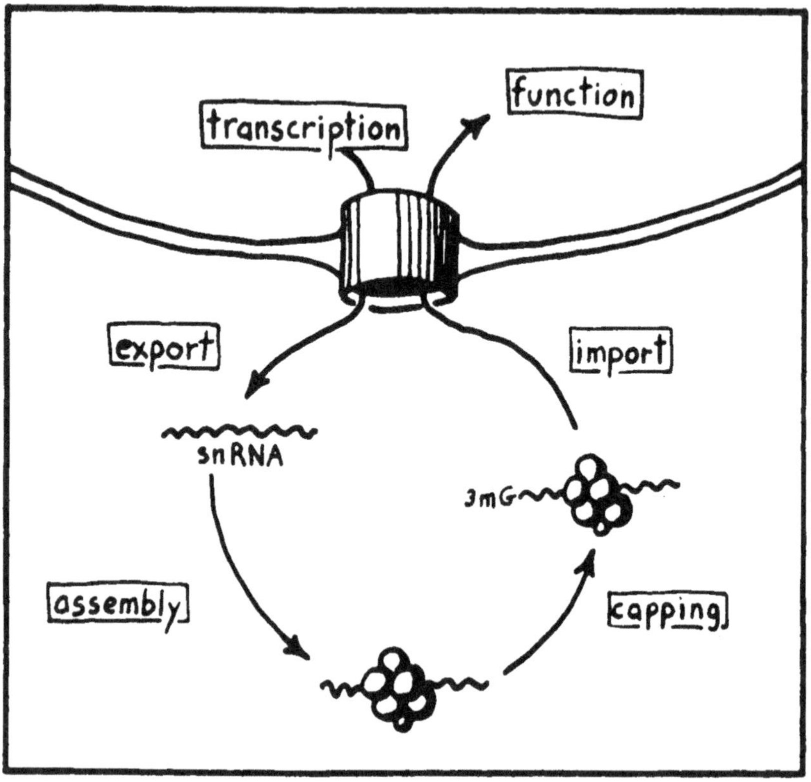

Figure 6. Nuclear trafficking during the biogenesis of U1 snRNP. U1 snRNA is transcribed in the nucleus, but is exported to the cytoplasm and assembled into a mature snRNP before it is imported back to the nucleus where it functions in pre-mRNA splicing.

inhibited by injecting high concentrations of a two-nucleotide long TMG cap analog. Future studies will focus on the identification of the transport factors unique to U snRNP import. How many other specialized import pathways exist is open to speculation.

The Export of Macromolecules from the Nucleus

In addition to importing nuclear proteins, the nucleus also synthesizes a number of RNAs and RNPs which must be exported to the cytoplasm.

A variety of messenger RNAs (mRNAs), ribosomal RNAs (rRNAs), and transfer RNAs (tRNAs) must exit the NPC to play their role in protein synthesis in the cytoplasm, a situation exactly opposite to that faced by karyophiles (Izaurralde and Mattaj, 1992). RNA-protein complexes represent the vast majority of export events, and it is thought that virtually all RNAs exit the nucleus in complexes with proteins.

Synthesis, Processing, and Export are Interconnected

The study of nuclear export is complicated by the fact that many export substrates must first be modified or processed in some way prior to transport. Thus, some control must be exerted on export to insure that immature versions of these RNAs are not allowed to exit the nucleus and interfere with cytoplasmic functions. mRNAs are synthesized as long intron-containing transcripts which must be spliced, tRNAs are spliced and processed at the 5' and 3' ends, and rRNAs are assembled into ribosomal subunits prior to export. In a normal cell, unprocessed RNAs are never seen in the cytoplasm, implying a high degree of control in this process. As we will see in the next section, some viruses (including HIV-1) have evolved techniques to avoid RNA processing and promote the export of unspliced viral RNAs.

Nuclear export may be thought of as a two-step procedure, in which RNAs are retained in the nucleus until posttranscriptional processing is complete, and then permitted or even encouraged to exit. This chain of interconnected events makes the study of nuclear export rather complicated, and even the most basic steps of export are uncertain. For example, one hypothesis states that processing events serve to retain RNAs in the nucleus via binding, and that only mature RNAs are truly free to move about the nucleus. Taken a step further, one could postulate that export is rather promiscuous, and any substrate freely soluble will be exported. Support for this idea came from experiments in which gold balls coated with a nonsense (noncoding) mRNA or DNA were exported as efficiently as actual messenger RNAs. These results suggest that there is no specificity in the export reaction.

Many Export Substrates are Actively Transported

Nonetheless, it is clear that at least some nuclear export is facilitated. In the studies just mentioned, the gold balls exceeded the diffusion limits of the pore, implying a requirement for facilitated transport of some sort. Indeed, many endogenous substrates are also too large to diffuse out of the nucleus. Finally, export may not be completely undiscriminating, since BSA and IgG, two cytoplasmic proteins, cannot exit the nucleus when microinjected.

The strongest evidence for active transport of RNPs to the cytoplasm comes from studies of mature mRNA, tRNA, and ribosomal subunits. tRNA export is saturable and temperature dependent, characteristic of carrier mediated transport. In the case of mRNA and ribosomal subunits, export is sensitive to ATP depletion and to WGA, an agent which interferes with pore functioning. Thus, it appears that export of RNAs is mediated much like import of proteins.

This leads to the question, "What are the signals for export?" Unfortunately, this question has only recently been studied, and little is known about elements that may function as export signals. Strictly speaking, the export signal may be some characteristic of the RNA species, or of an RNP complex. Current evidence points most strongly to the role of RNA in targeting. Studies which have examined the export of tRNAs containing various mutations indicate that single base changes in the sequence of the tRNA can have drastic effects on export. However, these changes may also interfere with processing or folding of the tRNA. mRNAs are synthesized with a monomethylguanosine cap, and this cap seems to play a role in directing export. It has been shown that the efflux of mRNA molecules can be inhibited by coinjection of large quantities of a dinucleotide cap, which apparently competes for some export factor. In the case of ribosome export, the complexity of the subunits has thus far precluded a systematic study of export requirements. However, it is known that the fidelity of the rRNA is important for efficient transport.

SHUTTLING PROTEINS

Although most export substrates are RNAs or RNPs, there also is a class of proteins which can traverse the nuclear envelope in both directions, and so the proteins have been dubbed "shuttling proteins" (Garcia-Bustos, et al., 1991; Forbes, 1992). Examples include nucleolin (a nucleolar protein), hsp70, and members of the steroid hormone receptor family. The significance of the shuttling phenomenon is unclear. Because of this apparently unique characteristic, it was originally thought that these proteins may be molecules whose function was to assist in the transport of proteins and RNAs as transport receptors.

For example, heterogeneous nuclear ribonucleoprotein (hnRNP) proteins bind to mRNA transcripts in the nucleus and then escort them to the cytoplasm. In the cytoplasm they are replaced by cytoplasmic proteins, and the hnRNP proteins shuttle back into the nucleus (Pinol-Roma and Dreyfuss, 1992). The shuttling behavior of some of these polypeptides was demonstrated in heterokaryons. A heterokaryon is a cell containing two or more nuclei, created by fusing cells together. If a protein contained in one nucleus shuttles, it will enter the cytoplasm and then reenter not only the original nucleus it inhabited, but other available nuclei as well. Thus, by observing the redistribution of nuclear proteins in heterokaryons, one can identify shuttling proteins. The identification of proteins which accompany mRNA out of the nucleus suggested that shuttling proteins participate in translocation events.

However, a recent analysis of several shuttling proteins of unknown function has painted a different picture. Export of these proteins proceeds extremely slowly, and in the case of the progesterone receptor, is temperature independent, implying a passive movement through the pore. The export characteristics of several proteins were examined, revealing that a protein does not require a positive signal to promote efflux. Rather, the main factor in determining whether a karyophile will shuttle is whether it has binding sites in the nucleus. If a protein has binding sites in the nucleus, it is retained. If a protein does not have binding sites, or if those binding sites are already filled, it is able to exit the nucleus.

The picture of export events that is emerging is a complex and intriguing one. Clearly, much remains to be discovered about export events and the coordination of processing and export.

REGULATED NUCLEAR TRANSPORT

Nuclear trafficking is regulated from the earliest stages of life. During early development, a variety of carefully orchestrated changes in cell regulation must take place to enact a developmental program. A comparison of the nucleocytoplasmic distribution of various proteins during early stages of development indicates that a number of proteins change their subcellular localization as development proceeds. Commonly, proteins are stored in the cytoplasm of unfertilized eggs and become translocated to the nucleus once development is initiated. This has been observed in organisms ranging from *Xenopus* to *Drosophila*, suggesting it is a common mechanism for cell control during embryogenesis. Indeed, in cases where individual proteins have been identified, they are often proteins thought to be involved in the regulation of specific events in development. Examples include c-myc, which controls cell proliferation, and FGF, which is thought to play a role in mesoderm induction.

A shift in the nucleocytoplasmic distribution of specific proteins can have dire consequences for a mature cell as well. abl is a proto-oncogene whose viral form is capable of transforming cells to an oncogenic phenotype. The viral form of abl, v-abl, is located in the cytoplasm, concentrated in the plasma membrane. However, the cellular counterpart, c-abl, is mostly nuclear in distribution. If a small portion of the N-terminus of c-abl is deleted, the protein's transforming ability is activated, and its distribution shifts from nuclear to cytoplasmic. This region of the abl gene contains an NLS similar to the SV40 T antigen signal. Thus, this protein functions properly when found in the nucleus, but inappropriate localization in the cytoplasm wreaks havoc on the cell and induces malignant growth.

The oncogene fos is an example of a gene product which functions as a master switch in signal transduction and the regulation of cell proliferation. The activation of fos induces a cascade of regulatory events postulated to be responsible for long-term cellular responses such as growth and differentiation. fos is a nuclear protein that associates with members of the jun-AP-1 family of transcription activators. In quiescent cells, c-fos is cytoplasmic, but when cells are stimulated with serum growth factors, c-fos rapidly accumulates in the nucleus. The viral form of the protein, v-fos, causes runaway cell growth, and is constitutively expressed and always localized in the nucleus. The nuclear location of v-fos appears to contribute to its tumorigenic potential.

NLS MASKING

As described above, the nuclear transport of most macromolecules is a multi-step process involving many cellular factors including NLS receptors and the NPC. Because the regulation of transport can potentially occur at any step along this pathway, it is possible to imagine multiple mechanisms of regulation (Figure 7):

1. The NLS may be masked in some fashion so that it cannot interact with transport receptors. The protein remains cytoplasmic because its targeting signal is inaccessible to the transport apparatus.

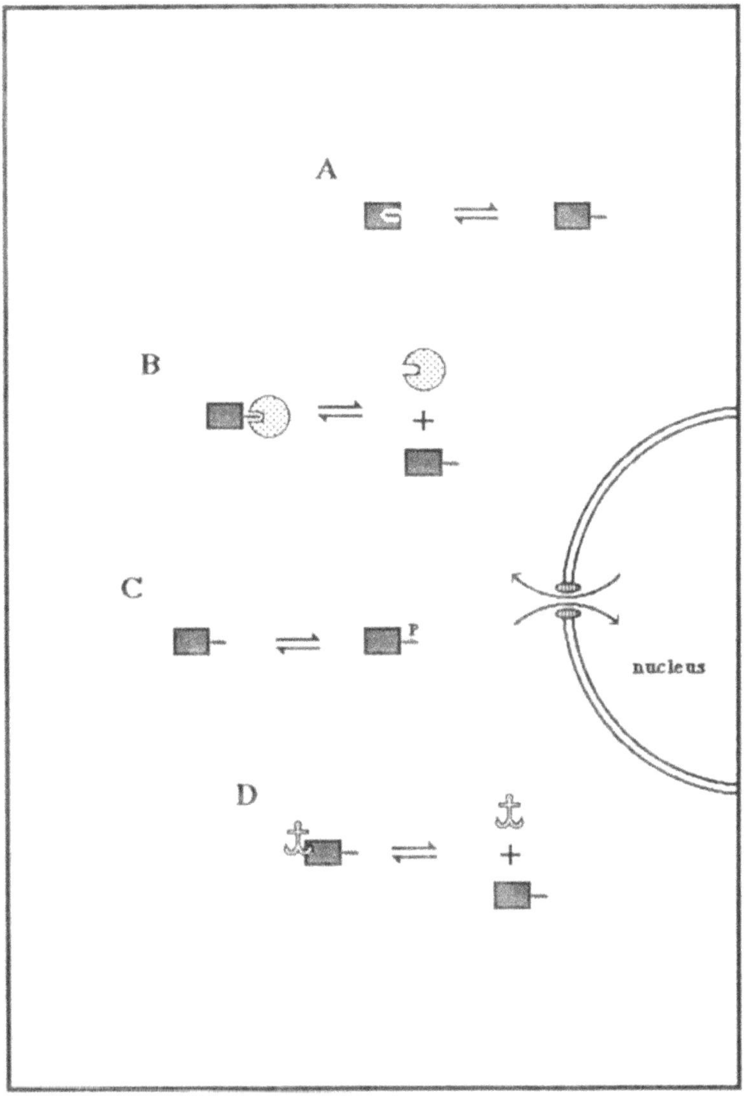

Figure 7. Possible mechanisms regulating nuclear transport. (**A**), (**B**) Examples of NLS masking. In (**A**), the conformation of the protein is such that the NLS is not exposed. An NLS can also be masked by the binding of a regulatory protein, as in (**B**). (**C**) Phosphorylation of amino acids surrounding an NLS can greatly increase the rate of nuclear transport. (**D**) Nuclear proteins sequestered in the cytoplasm through interactions with immobilized proteins are unable to move to the nuclear pore complex.

2. A nuclear protein may be physically anchored in the cytoplasm, and thus unable to move into the nucleus.
3. The permeability of the NPC itself may be regulated.

There is accumulating evidence that regulation of nuclear transport does occur at all levels. Interestingly, the proteins involved often play key roles in cell regulation and proliferation, two processes that often go awry in many diseases.

First, we will review cases in which the NLS of a protein is inaccessible to a transport factor or receptor. The masking of an NLS can be accomplished in several ways. In some cases, a regulatory protein binds the NLS of a karyophile, preventing transport receptors from interacting with the signal. It is also possible that cytoplasmic forms of otherwise nuclear proteins are present in structural conformations that sterically encrypt the NLS (Hunt, 1989).

NF-kB represents a class of transcription factors whose nuclear localization is regulated. NF-kB acts to control the transcription of immunoglobulin genes in the nucleus, but is commonly found in an inactive form in the cytoplasm. The nuclear import of NF-kB can be stimulated by a variety of agents which are thought to be involved in signal transduction. In the cytoplasm, NF-kB binds an inhibitory protein called IkB. When NF-kB is activated, it is released from the IkB and can move into the nucleus (Henkel et al., 1992). IkB is thought to inhibit NF-kB by binding to its NLS, rendering it inaccessible to the transport machinery.

The DNA-binding subunit of NF-kB has homologies to two other proteins known to participate in cell regulation: dorsal, a *Drosophila* protein involved in dorsal-ventral patterning in the early embryo and rel, an oncogene involved in neoplastic transformation. In *Drosophila* embryos, which at "precellular" stages of development contain hundreds of nuclei in a single cytoplasm, the activity of the dorsal protein depends on its subcellular localization. The protein is found throughout the precellularized embryo, but on the dorsal side of the embryo it is cytoplasmic, while on the ventral side of the embryo, it is nuclear (see Rushlow et al., 1989). The asymmetric nuclear localization of dorsal protein is crucial for proper development, since nuclear dorsal acts to induce the transcription of genes that direct the production of ventral structures in this region of the embryo. Little is known about the proteins which participate in the regulated nuclear localization of dorsal, although recently a gene interacting with dorsal was found to have similarities to IkB. There are some clues as to how the distribution of dorsal is regulated. When the dorsal protein is expressed in tissue culture cells, it is cytoplasmic. If the same cells are transfected with dorsal protein which lacks eight amino acid residues from its carboxy terminus, the protein becomes nuclear. This suggests that some type of NLS masking occurs in the cytoplasmic form of the protein.

rel is a proto-oncogene whose cellular homologue is cytoplasmic. However, the viral gene product, v-rel, is nuclear in some cell lines. A carboxy-terminal sequence is required for cytoplasmic localization, a situation analogous to the dorsal protein.

REGULATION BY PHOSPHORYLATION

A major mechanism for the regulation of protein activity is phosphorylation (Forbes, 1992; Hunter and Hall, 1992). Phosphorylation and dephosphorylation control a host of cell activities, and there is evidence that this may be a significant means of regulating the nuclear localization of many proteins. This work began with the early observation that a short amino acid sequence of the SV40 T antigen was necessary and sufficient for nuclear import of this protein. However, if sequences flanking either side of the NLS were included, transport was greatly enhanced (Jans et al., 1991). A careful examination of the flanking sequences revealed that they contained sites for phosphorylation of serine and threonine residues. Further, one of the threonine sites was known to be phosphorylated *in vivo*. The import of SV40 T antigen fusion proteins containing phosphorylated and unphosphorylated threonine residues was compared. Fusion proteins phosphorylated on threonine had greatly accelerated rates of import. Interestingly, an examination of some other nuclear proteins has revealed that they, too, possess phosphorylation sites in or near their NLSs. In the case of one yeast protein, SWI 5, phosphorylation regulates the localization of the protein during specific phases of the cell cycle. SWI 5 phosphorylation is accomplished by CDC 28 kinase, a protein implicated in controlling the cell cycle in yeast.

CYTOPLASMIC TETHERING OF NUCLEAR PROTEINS

Protein access to the nucleus can also be regulated by anchoring the protein in the cytoplasm. The regulation of the cAMP-dependent protein kinase is an excellent example of a nuclear protein which can be denied access to the nucleus by cytoplasmic anchoring. cAMP-dependent protein kinase plays a vital role in the transduction of signals from the cell surface to the nucleus. The kinase is composed of regulatory and catalytic subunits, and when the regulatory and catalytic subunits are complexed, the protein is inactive and restricted to the cytoplasm. An extracellular signal received at the plasma membrane initiates a cascade of events resulting in the stimulation of a second messenger system. Cyclic AMP (cAMP) is one such messenger. When cAMP binds the protein kinase complex, the regulatory and catalytic units dissociate and the catalytic subunits become localized to the nucleus while the regulatory subunits remain cytoplasmic. Work by Nigg and colleagues suggests that the regulatory subunits are bound to the membrane of the Golgi apparatus and are unable to move freely through the cytoplasm (Nigg et al., 1985). Thus, when catalytic units are bound to regulatory units they are effectively tethered in the cytoplasm.

REGULATION OF STEROID HORMONE RECEPTORS

At first glance, steroid hormone receptors would appear to be a class of proteins ideally suited to regulation of nuclear import. Steroid hormone receptors are transcription factors which require ligand binding for activation. The hormone receptor complex is then able to bind DNA promoter elements in the nucleus and regulate the transcription of a variety of genes. Early models for the functioning of this complex postulated that receptors remained cytoplasmic and inactive until hormone binding induced translocation to the nucleus. However, recent studies indicate that most hormone receptors are able to enter the nucleus even in an unbound, and therefore inactive, state.

Nonetheless, at least one receptor, the glucocorticoid receptor, requires ligand binding for rapid translocation to the nucleus (Picard and Yamamoto, 1987). The glucocorticoid receptor contains two NLSs, one of which resides in the ligand binding domain of the protein. It is this second NLS which appears to be responsible for regulating nuclear import of the complex. While little is known about how hormone binding drives nuclear localization of the receptor, one model suggests that binding of hormone unmasks a NLS contained in the hormone-binding domain. For other hormone receptors the question remains: How can the nuclear import of unbound, inactive receptors allow efficient regulation of transcription?

Work done on the progesterone receptor has yielded several clues to this mystery. Like many steroid hormone receptors, the progesterone receptor contains two NLSs. One is constitutive, but when this signal is deleted, the protein can still enter the nucleus in a hormone-dependent fashion. Further, nuclear import of the receptor requires energy, but the receptor can exit the nucleus freely, apparently by simple diffusion. Thus, the receptor continually shuttles back and forth between the nucleus and cytoplasm, so that receptor is always available to bind hormone in the cytoplasm. It will be interesting to see if the rate of nuclear import of the receptor increases upon binding hormone.

The heat shock protein hsp90 may also play a role in regulating the nucleocytoplasmic distribution of steroid hormone receptors. *In vitro*, steroid hormone receptors can bind DNA elements without binding hormone or becoming activated. However, *in vivo*, ligand binding is required for DNA binding. In the cell, steroid hormone receptors are found in the cytoplasm in a complex with hsp90 (and perhaps other proteins). Thus, it has been suggested that hsp90 functions to anchor a receptor in the cytoplasm until it is displaced by the binding of hormone. Evidence for this model comes from experiments in which cells are treated with compounds that cause the dissociation of hsp90 from receptor. These freed receptors are now competent to translocate into the nucleus.

Finally, it is possible to imagine that nuclear transport may be regulated at the NPC itself. Unfortunately, at this time, almost nothing is known about how the NPC functions to translocate proteins. However, studies done on the transport of various substrates suggests that the permeability of the NPC may vary under different physiological conditions and for different transport substrates. For example, dividing cells transport larger karyophiles more efficiently than non-dividing cells (Feldherr and Akin, 1991). In proliferating cells, the functional size of the NPC pore is 26 nm, but in confluent cells, this decreases to about 11 nm. Other researchers have shown that the efficiency of nuclear transport is maximized in the presence of peptide growth factors, and in transformed cells. These conditions affect bidirectional nuclear transport, including both diffusion and active transport. Obviously, these changes affect nuclear transport as a whole, but take on added significance in light of the observation that many viral particles which gain entry to the nucleus are significantly larger than endogenous proteins. These observations and others discussed below may eventually be parlayed into strategies for prevention of viral infection.

NUCLEAR TRAFFICKING OF VIRUSES

Many viruses must spend part of their life cycle in the host nucleus, exploiting the nuclear machinery of their host to replicate their genomes and transcribe viral genes (see Table 2). Some examples include adenovirus, influenza, and, of course, HIV-1. A variety of mechanisms for the nuclear entry of viruses have been proposed, most based on morphological observations of viruses within host cells. For example, nothing is known about the mechanism by which adenovirus transfers its genome to the nucleus, although several possibilities have been suggested on the basis of electron micrographs of infected cells. Adenovirus may fuse with the nuclear envelope and expel its genetic material into the nucleus, or the virus particle may cross the NPC by the same mechanism used by endogenous karyophiles. While these models are intriguing, there is currently no strong evidence to support either model. Research on the nuclear import of other viruses has been more revealing, and in the case of SV40 and influenza, it is clear that the viral particles use a mediated transport which may be similar to other karyophiles. Elucidation of the import mechanisms may lead to strategies for prevention of viral infection and proliferation.

The most well studied incidence of viral nuclear transport is the influenza A virus (Martin and Helenius, 1991). During influenza A infection, the virus is taken up in the cell by endocytosis. Once in the host, the virus particle is released into the cytosol following a pH-activated fusion event. This causes proteins of the viral capsid to dissociate, releasing the enclosed viral genome. The influenza genome is comprised of several viral RNA-protein complexes (vRNPs). These vRNPs enter the nucleus where new RNAs are transcribed. The import process appears to be energy dependent, and utilizes the NPC, much like the import of endogenous proteins. Indeed, the major protein of the vRNPs, nucleoprotein, contains an NLS similar to those found in other karyophiles.

Interestingly, the timing of import and export of these vRNPs appears to be controlled in a unique fashion. Upon entry to the cell, the vRNPs become dissociated from a viral protein called M1, and this dissociation is required for entry into the nucleus. New vRNPs are synthesized in the nucleus and require M1 protein for export. Thus M1 functions as a regulator of both import and export of the vRNPs. This situation is remarkably analogous to mechanisms of regulation utilized by endogenous nuclear proteins.

Regulation of Nuclear Trafficking by Viral Proteins

There are several instances of viral proteins which appear to control the nuclear trafficking of RNAs for their own benefit. This usually involves inhibiting the export of host mRNAs while encouraging the export of virus-encoded mRNAs. HIV-1 infection requires the expression of two sets of viral RNAs. The early RNAs are transcribed in the host nucleus, spliced, and transported to the cytoplasm where they encode regulatory proteins. The second set of transcripts are incompletely spliced late RNAs which encode viral structural proteins. The host cell normally controls the export of mRNA so that RNAs which are not completely spliced are retained in the nucleus until processing is completed. However, the late RNAs from HIV are exported to the cytoplasm despite the fact that they are not fully spliced. This export event appears to be mediated by a gene product of the early RNAs, Rev. Rev does not appear to affect the rate of RNA synthesis or the stability of the RNAs, but actually seems to function as a catalyst for export of these viral RNAs (Malim et al., 1989). An export element located in the introns of the viral RNAs functions to mediate the export event. The element is rather large and has the potential to form extensive secondary structure, suggesting that it directly binds some regulator protein, possibly Rev itself.

An analogous situation is reported for adenovirus. In cells infected with adenovirus, a precise program of transcription occurs in the nucleus of the host cell. A set of early genes is transcribed first, including the E1B gene, which encodes a protein that regulates the cytoplasmic accumulation of late vRNAs and host mRNAs. E1B protein allows the accumulation of the late viral RNAs in the cell cytoplasm, while at the same time inhibiting the cytoplasmic accumulation of host RNAs. The function of this protein has been studied using adenovirus mutants which are deleted for the E1B gene. In these mutants, the late genes are transcribed and processed normally, suggesting that E1B exerts its control post-transcriptionally. Further, in these mutants, only small amounts of the late RNAs are found in the cytoplasm, suggesting that the E1B protein is involved in facilitating the export of late viral RNAs from the nucleus (Leppard and Shenk, 1989). Other adenovirus genes act to inhibit the expression of host genes, resulting in the hijacking of the cell's protein synthesis apparatus for the production of viruses. This dramatic strategy results in the production of many viruses, but ultimately kills the host cell.

SUMMARY

The nuclear envelope separates the nucleus from the cytoplasm. Consisting of two membranes and the nuclear lamina, the nuclear envelope is penetrated by nuclear pore complexes (NPC) which are the sites of macromolecular transport between the nucleus and cytoplasm. The nuclear envelope undergoes a cycle of disassembly and reassembly during each mitosis. During mitosis in higher eukaryotes, many nuclear components mix with the cytoplasm and must, therefore, be re-sorted following the reassembly of the daughter cell nuclear envelopes.

Macromolecules destined for the nucleus contain nuclear localization signals (NLSs) that direct them to the NPC via interactions with NLS-receptors. The translocation of macromolecules across the nuclear envelope requires ATP hydrolysis, although how this energy is transduced into vectorial transport is unknown. The control of RNA export is exquisite, as only fully spliced and processed RNAs are allowed to exit.

The transport of some macromolecules is regulated during the cell cycle, development, changing environmental conditions, and disease. Mechanisms of regulated nuclear transport include the phosphorylation of sequences adjacent to the NLS, the masking of NLSs, and the tethering of molecules to cytoplasmic substrates. During many viral infections the expression of host genes is down-regulated in favor of viral gene expression, for example, by preferentially inhibiting the export of host mRNAs.

During their life cycle, many viruses, such as influenza A and adenovirus, enter interphase nuclei via the NPC. In contrast, most retroviruses enter the nucleus only during mitosis when the integrity of the nuclear envelope is compromised. These viruses cannot, therefore, infect nondividing cells. Like influenza A and adenoviruses, the lentiviruses, such as HIV-1, contain NLSs that direct their nuclear transport. Therefore, unlike most retroviruses, the lentiviruses are able to infect nonproliferating cells of the nervous and immune systems.

REFERENCES

Adam, S.A., & Gerace, L. (1991). Cytosolic proteins that specifically bind nuclear location signals are receptors for nuclear import. Cell 66, 837-847.

Akey, C. W. & Radermacher, M. (1993). Architecture of the Xenopus nuclear pore complex revealed by three-dimensional cryo-electron microscopy. J. Cell Biol. 122, 1-19.

Breeuwer, M., & Goldfarb, D.S. (1990). Facilitated nuclear transport of histone H1 and other nucleophilic proteins. Cell, 60, 999-1008.

Bukrinsky, M., Haggerty, S., Dempsey, M., Sharova, N., Adzhubei, A, Spitz, L., Lewis, P., Goldfarb, D., Emerman, M., & Stevenson, M. (1993). A nuclear localization signal within HIV-1 matrix protein that governs infection of non-dividing cells. Nature 365, 666-669.

Dingwall, C., Sharnick, S.V., & Lasky, R.A. (1982). A polypeptide domain that specifies migration of nucleoplasmin into the nucleus. Cell 30, 449-453.

Feldherr, C.M. & Akin, D. (1991). Signal-mediated nuclear transport in proliferating and growth-arrested BALB/c 3TC cells. J. Cell Biol. 115, 933-939.

Feldherr, C.M., Kallenbach, E. & Schultz, N. (1984). Movement of karyophilic proteins through the nuclear pores of oocytes. J. Cell Biol. 99, 2216-2222.

Fields Virology. Vol. 2, Second Ed. (Fields, B.N. & Knipe, D.M., Eds.) Raven Press, New York.

Forbes, D. J. (1992). Structure and function of the nuclear pore. Ann. Rev. Cell Biol. 8, 495-528.

Garcia-Bustos, J., Heitman, J., & Hall, N.M. (1991). Nuclear protein localization. Biochim. Biophys. Acta 1071, 83-101.

Goldfarb, D. S. & Michaud, N. (1991). Pathways for the nuclear transport of proteins and RNAs. Trends Cell Biol. 1, 20-24.

Goldfarb, D. S. (1992). Are the cytosolic components of the nuclear, ER, and mitochondrial import apparatus functionally related? Cell 70, 185-188.

Henkel, T., Zabel, U., van Zee, K. Muller, J.M., Fanning, E., & Baeuerle, P.A. (1992). Intramolecular masking of the nuclear location signal and dimerization domain in the precursor for the NK-kB subunit. Cell 68, 1221-1233.

Hinshaw, J.E., Carragher, B.O., & Milligan, R.A. (1992). Architecture and design of the nuclear pore complex. Cell 69, 1133-1141.

Hunt, T. (1989). Cytoplasmic anchoring proteins and the control of nuclear localization. Cell 59, 949-951.

Hunter, T. & Hall, A. (1992). The regulation of transcription by phosphorylation. Cell 70, 375-388.

Izaurralde, E. & Mattaj, I. W. (1992). Transport of RNA between nucleus and cytoplasm. Semin. Cell Biol. 3, 279-288.

Jans, D.A., Ackermann, M.J., Bischoff, J.R., Beach, D.H., & Peters, R. (1991). $p34^{cdc2}$-mediated phosphorylation at T^{124} inhibits nuclear import of SV-40 T antigen proteins. J. Cell Biol. 115, 1203-1212.

Laskey, R. A. & Leno, A. (1990). Assembly of the cell nucleus. Trends. Genet. 6, 406-410.

Leppard, K.N. & Shenk, T. (1989). The adenovirus E1B 55 kd protein influences mRNA transport via an intranuclear effect on RNA metabolism. EMBO J. 8, 2329-2336.

Malim, M.H., Hauber, J., Le, S.-Y., Maizel, J.V., & Cullen, B.R. (1989). The HIV-1 *rev trans*-activator acts through a structured target sequence to activate nuclear export of unspliced viral mRNA. Nature 338, 254-257.

Martin, K. & Helenius, A. (1991). Nuclear transport of influenza virus ribonucleoproteins: The viral matrix protein (M1) promotes export and inhibits import. Cell 67, 117-130.

Moore, M.S. & Blobel, G. (1992). The two steps of nuclear import, targeting to the nuclear envelope and translocation through the nuclear pore, require different cytosolic factors. Cell 69, 939-950.

Nigg, E.A., H. Hilz, Eppenberger, H.M., & Dutly, F. (1985). Rapid and reversible translocation of the catalytic subunit of cAMP-dependent protein kinase type II from the Golgi complex to the nucleus. EMBO J. 4, 2801-2806.

Osborne, M. A. & Silver, P. A. (1993). Nucleocytoplasmic transport in the yeast Saccharomyces cerevisiae. Ann. Rev. Biochem. 62, 219-254.

Paine, P.L., Moore, L.C., & Horowitz, S.B. (1975). Nuclear envelope permeability. Nature 254, 109-114.

Picard, D. & Yamamoto, K.R. (1987). Two signals mediate hormone-dependent nuclear localization of the glucocorticoid receptor. EMBO J. 6,3333-3340.

Pinol-Roma, S. & Dreyfuss, G. (1992). Shuttling of pre-mRNA binding proteins between the nucleus and cytoplasm. Nature 355, 730-732.

Rushlow, C.A., Han, K., Manley, J.L., & Levine, M. (1989). The graded distribution of the *dorsal* morphogen is initiated by selective nuclear transport in Drosophila. Cell 59, 1165-1177.

Yamasaki, L. & Lanford, R. E. (1992). Nuclear localization signal receptors. Trends Cell Biol. 2, 123-127.

Chapter 3

Chromosomes, Chromatin, and the Regulation of Transcription

NICO STUURMAN and PAUL A. FISHER

Principles of Medical Biology, Volume 2
Cellular Organelles, pages 55-71

ISBN:1-55938-803-X

INTRODUCTION

To fit inside a cell nucleus, DNA must be compacted manyfold. Compaction into chromatin and chromosomes is achieved through protein binding. In living cells, DNA is never free but is always associated with proteins. Indeed, proteins constitute greater than half the mass of interphase chromatin, the predominant DNA-containing complex in the nondividing cell. At the first level, DNA is folded around an octamer of histones to form nucleosomes. Nucleosomes are organized into 30 nm fibers. At higher levels, molecular details of DNA folding are poorly understood. Cell differentiation and development of organisms result, at least to a degree, from differential activation and/or inactivation of specific genes. Current evidence suggests that chromatin structure is important in regulating gene expression. This is in conjunction with diffusible regulatory factors (proteins) and active enzymes such as RNA polymerases. More detailed understanding of chromatin and chromosome structure will be required to critically evaluate the roles they play in controlling transcription and, hence, normal differentiation and development.

OVERVIEW

DNA, the Hereditary Material, Is Folded Into Chromatin

Results of many experiments culminating in the elucidation of its double-helical structure (Watson and Crick, 1953), established the role of DNA as the molecular carrier of genetic information. Although considerable progress has been made toward understanding how this information is stored, expressed, and transmitted from one generation to the next, a number of fundamental questions, pertaining to each of these processes, remain. The notion that chromatin (chromosome) structure, resulting from the interaction of DNA and proteins, plays an important role in these processes, is now widely accepted (Wolffe, 1992; Felsenfeld, 1992). The potential importance of chromatin structure may be particularly well-appreciated when one considers that a typical mammalian nucleus with a diameter of about 5 microns contains more than 2 m of DNA packed within.

Chromatin Occurs in Many Forms, Is Ordered and Dynamic

Because of their distinctive morphology at the light microscopic level, condensed chromosomes, both mitotic and meiotic, were the first recog-

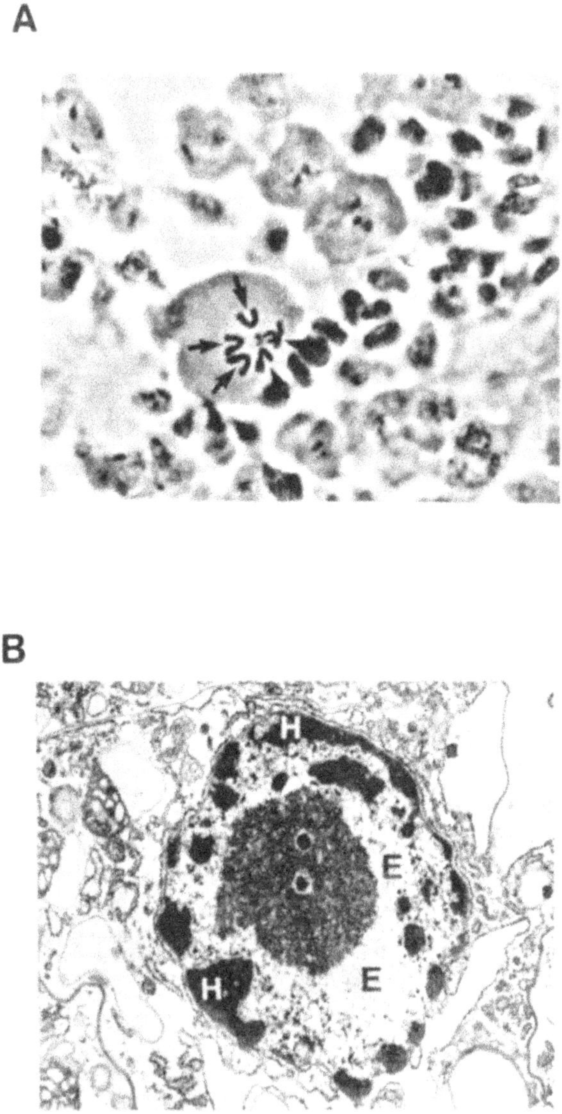

Figure 1. Chromosomes and chromatin. A, condensed chromosomes within a mitotic *Drosophila* third instar larval neural ganglion cell. Mitotic chromosomes, stained with orcein, are indicated by arrows. B, transmission electron micrograph of a *Drosophila* embryonic nucleus showing heterochromatin (H) and euchromatin (E) as indicated (micrograph courtesy of Dr. M. Berrios, Stony Brook).

nized forms of chromatin. Seen during somatic cell division, mitotic chromosomes (Figure 1A) consist of two sister chromatids, each destined for distribution to different daughter cells. Mitotic sister chromatids are joined at a central constriction termed the centromere; the ends of each chromosome are termed telomeres. Structure as well as function of telomeres and centromeres are discussed in Chapter 5.

Mitotic (meiotic) chromosomes can be histochemically stained in a number of ways resulting in species-specific and chromosome-specific banding patterns (Jeon and Friedlander, 1987). Each visible band is thought to contain many genes. Recognizable changes in banding patterns are indicative of major chromatin rearrangements which, when associated with specific genetic defects, can be of significant diagnostic value. Moreover, the very existence of chromosome bands is indicative of extended higher-order chromatin structure.

At the end of mitosis (telophase), i.e., the beginning of interphase, chromosomes decondense and, in most cases, become morphologically unrecognizable by conventional techniques. Coincidentally, a new nucleus surrounding decondensed chromatin forms in each daughter cell. Results of transmission electron microscopy indicate that during interphase, at least two forms of chromatin can be identified. Heterochromatin is morphologically distinguishable as being more condensed; euchromatin is relatively less condensed (Figure 1B). After visualizing transcription (RNA synthesis) with labeled RNA precursors, it became clear that most RNA synthesis takes place within areas of euchromatin, while heterochromatin is relatively inactive. Some portions of the genome, e.g., centromeres and telomeres, are found in heterochromatic regions of all cell types. These regions are termed constitutive heterochromatin. Other portions of the genome are found in heterochromatic regions of some cell types but not others. These regions are termed facultative heterochromatin. It seems likely that partitioning of specific genomic regions to facultative heterochromatin represents a mechanism for 'turning off' (silencing) large portions of chromosomes, the expression of which is not needed in a specific cell type.

Recently developed techniques allow visualization in certain cell types of complete individual chromosomes in interphase nuclei. As a result, it is now apparent that at least in these cells, individual chromosomes occupy distinct and separate domains (see Alberts et al., 1994, p. 383). This observation suggests that decondensed interphase chromosomes, like their condensed mitotic relatives, are highly ordered.

Chromatin Structure and Gene Expression

It is essential to all multicellular organisms that genetic material be expressed in a precisely-timed and highly-regulated fashion. This is particularly well-illustrated by embryogenesis. Pluripotent stem cells differentiate into virtually all of the specialized cell lineages of the organism with selective activation or preservation in an activatable state, of certain genes and more-or-less permanent silencing of many other genes. Gene expression is regulated on many levels. All genes are transcribed by RNA polymerases, multisubunit enzymes capable of producing specific RNA by copying a DNA template. At the first level, the activity of RNA polymerases is controlled by the action of specific transcription factors. Transcription factors are proteins which bind DNA in a sequence-specific manner and either facilitate (positive factors) or suppress (negative factors) specific local actions of RNA polymerases. At higher levels, it is clear that access of key regulatory DNA sequences to various transcription factors is highly dependent on chromatin structure. So too is access of genes themselves to RNA polymerases. For example, genes that are condensed into 30 nm fibers are thought to be completely inactive transcriptionally.

Recently, several advances have come about in our understanding of chromatin structure and many implications for regulation of gene expression have been recognized. To detail these advances and their implications would require at least a separate volume. Rather, we have chosen to focus on the highlights of current understanding in these areas. Models of chromatin structure will be summarized and examples of how chromatin structure is known to affect gene expression will be presented.

DNA FOLDING

Core histones and the nucleosome

Eukaryotic chromatin contains large amounts of a group of small basic (as opposed to acidic) proteins termed histones. The four core histones, 2A, 2B, 3, and 4 form an octamer composed of two molecules of each core histone. Double-stranded DNA is wrapped nearly twice around each histone octamer to form the typical nucleosome. Core histones each have an average mass of about 12 kilodaltons. The proteinaceous component of the nucleosome thus has a minimal mass of about 100 kilodaltons (kD). The DNA component of the nucleosome also has a mass of about 100 kD. Nucleosomal organization and histone composition are widely dispersed and highly conserved in nature. For example, histone 4 from the pea plant is virtually identical with its bovine homolog differing in only two amino acids out of a total of 102. Nevertheless, it should be noted that some lower eukaryotes, e.g., dinoflagellates, package their DNA in other ways; similarly, the specialized chromatin found in mature mammalian sperm is highly condensed through the action of a unique class of proteins termed protamines (as opposed to histones) (Wolffe, 1992).

Much has been learned about the general principles of chromatin structure by *in vitro* studies of isolated histones and DNA-containing nucleosomes. Nucleosomes can be disrupted by incubation in buffers of higher than physiological ionic strength. This observation suggests that electrostatic forces are essential for maintaining the histone-histone and/or histone-DNA interactions required for structural stability of the nucleosome. In the absence of DNA, histones 3 and 4 form a tetramer composed of two molecules each of each histone. Histones 2A and 2B form a heterodimer. When histones are mixed with free DNA in solutions of high ionic strength and the ionic strength is then gradually reduced, the histone 3/4 tetramer binds first and more tightly to DNA, followed thereafter by two histone 2A/2B dimers. In this way, the nucleosomal structure of chromatin can be reconstituted *in vitro*. The mechanism of *in vitro* reconstitution, i.e., initial tight binding of the histone 3/4 tetramer followed by looser binding of two histone 2A/2B dimers, is thought to mimic the mechanism of chromatin assembly in living cells.

Cloning and sequencing of many histone genes has already been accomplished. Deduced amino acid sequences confirm the small size and basic nature of these proteins. Overall, greater than 20 % of the amino acids are basic (lysine or arginine) and thus, positively charged at physiological pH. Computer predictions of structure as well as results of X-ray diffraction studies of histones in nucleosomes suggest that all four core histones contain a globular COOH-terminal domain which mediates both histone-histone and histone-DNA interactions. The NH_2-terminal domain of histones contains most of the lysine and arginine residues and is the primary site for posttranslational modification by phosphorylation and acetylation.

In the octamer, the globular domains of core histones form a disk of about 11 nm in diameter and 5.5 nm in height. It is thought that histone

A

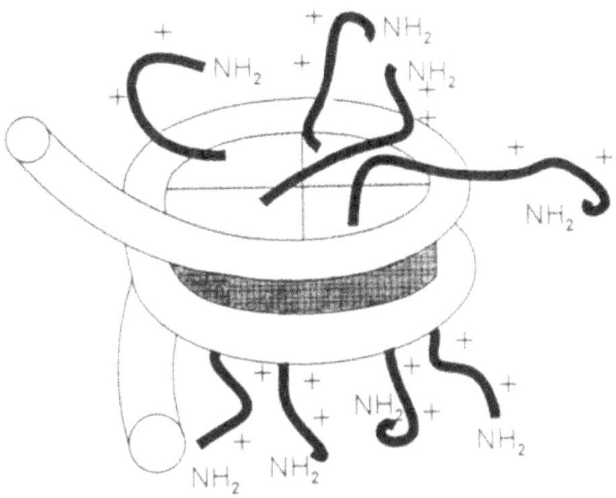

B

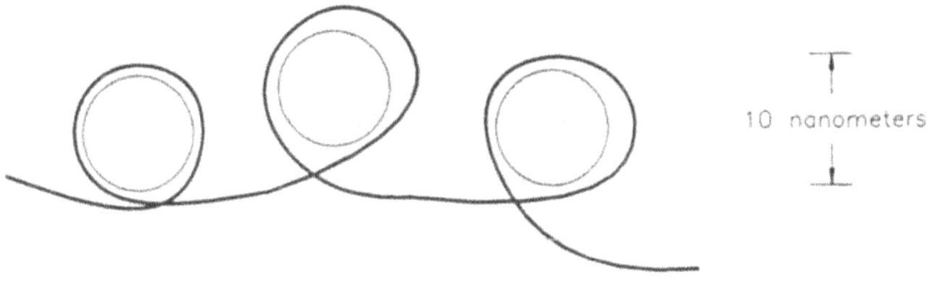

Figure 2. Chromatin structure. A, structure of the nucleosome (adapted from Grunstein, 1992). B, the 10-nanometer filament showing "beads on a string" (adapted from Alberts et al., 1994; and Grunstein, 1992).

NH_2-terminal tails extend from this disk and are accessible for binding to other proteins or DNA. In the nucleosome (Figure 2A), 146 base-pairs of DNA duplex form 1.75 turns around each disk such that DNA enters and leaves the nucleosome at about the same position. Nucleosomal DNA is of the typical B-configuration with an average of 10.2 base-pairs per turn of a left-handed double helix. This calculates to a total of about 14.3 turns of the helix per nucleosome. When free in solution, double-helical DNA contains 10.5 base-pairs per turn such that there is a net reduction in the number of base-pairs per turn when free DNA is assembled into·nucleosomes. Moreover, within nucleosomes, there is local distortion of the DNA structure. The central three turns of the double helix contain about 10.7 base-pairs per turn, whereas the remaining DNA exhibits about 10.0 base-pairs per turn. Local distortions of DNA structure are thought to have functional implications for nucleosome positioning. The nucleosomal organization of chromatin can be visualized by transmission electron microscopy. Under appropriate conditions, isolated chromatin appears as beads on a string (Figure 2B); also known as 10 nanometer filaments.

Histone H1 and the 30 Nanometer Filament

Nucleosomes are connected to each other by short DNA segments termed linkers (see Figure 2 B). A fifth histone, histone H1 (histone H5 in avian erythrocytes) binds specifically to linker DNA. Although basic like the core histones, histone H1 is much larger in molecular mass and does not participate directly in nucleosome formation. Rather, the globular central part of histone H1 binds core DNA at the points of entry to and exit from nucleosomes, thereby 'sealing' and thus stabilizing nucleosome structure. The long, highly basic COOH-terminal domain of histone Hl binds directly to the acidic phosphate backbone of linker DNA and promotes further folding of chromatin from 10 nanometer filaments to 30 nanometer fibers (see Alberts et al., 1994, p. 345).

Although folding from the 10 nm filament to the 30 nm fibers is readily accomplished *in vitro*, molecular details of the folded structure are still not fully understood. At least one explicit model (Felsenfeld and McGhee, 1986) has been articulated. This model proposes that 10 nm filaments fold into simple solenoidal structures with about six nucleosomes per helical turn. The helical solenoid has a pitch of about 11 nm per turn and linker DNA is believed to be coiled and located along the outside of the fibers (see Felsenfeld and McGhee, 1986).

The Looped Domain Model of Higher-Order Chromatin Structure

Folding of DNA first into 10 nm filaments and then into 30 nm fibers results in about a 300-fold reduction in DNA length. To fit within a typical mammalian nucleus, DNA length must be reduced at least 10,000 fold. Clearly, some higher-order packing scheme or schemes are operative. However, despite numerous proposals, the essential concepts involved remain highly controversial. Several pertinent ideas have resulted from studies of a few specialized forms of chromatin.

Certain insect secretory tissues such as the salivary gland of the fruit fly (*Drosophila melanogaster*) normally produce extraordinarily large amounts of protein. To facilitate protein production, *Drosophila* salivary gland cells amplify virtually their entire genomes by a process known as chromosome polytenization. Polytenization is due to multiple rounds of DNA replication without cell division or separation of sister chromatids. As a result, polytene chromosomes contain as many as 1,000 copies of each chromosome aligned in register with each other. Although otherwise a typical interphase chromosome, polytene chromosomes are visible by conventional light microscopy and exhibit a distinctive banding pattern (Figure 3). Unlike the much larger bands of mitotic chromosomes, each polytene chromosome band is thought to represent only a single-transcription unit, i.e., a single gene. When specific genes are activated (e.g., during heat shock exposure of *Drosophila* to elevated temperatures), individual bands containing these genes change morphologically such that condensed chromatin expands into decondensed loops and specific messenger RNAs (mRNAs) are transcribed from the decondensed chromatin. A similar looping pattern is observed in the so-called lampbrush chromosomes of amphibian oocytes. Enormous amounts of RNA are rapidly produced from these single chromosomes and, as a result, chromatin loops containing specific transcription complexes and transcribed RNA can be visualized (see Alberts et al., 1994, p. 347). Based upon observations of polytene and lampbrush chromosomes as well as results of other experiments, it was proposed that in nuclei, chromatin fibers are organized into loops of 50-100 kilobases. In support of this notion, it was observed that mitotic chromosomes from which histones were carefully removed chemically apparently remained attached to a central scaffolding structure which served to anchor DNA loops (Figure 4). Similarly, an interphase chromosome scaffold (also known as an internal nuclear matrix) has been found after chemical removal of histones from nuclei. Although potentially of profound significance in

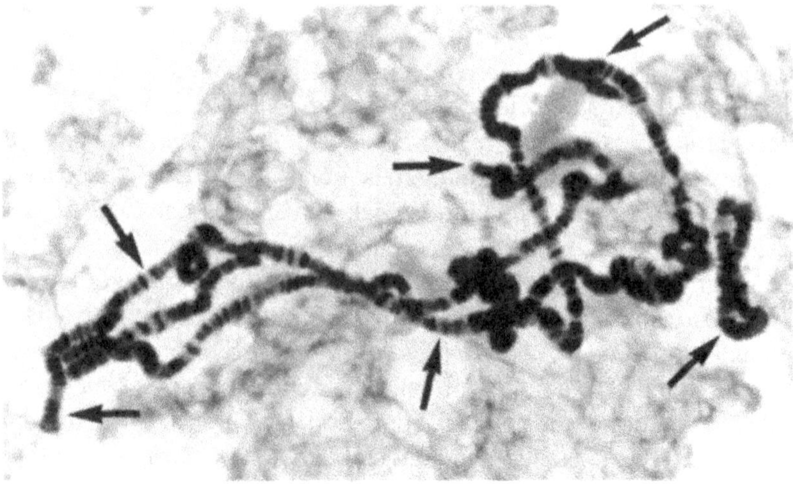

Figure 3. Polytenized interphase chromosomes. *Drosophila* third instar larval salivary glands were gently squashed and stained with orcein. Polytene chromosomes are indicated by arrows and are derived from a single nucleus.

organizing DNA within nuclei (Gasser and Laemmli, 1987; Goldman, 1988; van Driel et al., 1991), the very existence of such a scaffold remains controversial. This is due, at least in part, to the many artifacts associated with cell fractionation (Fisher, 1989), coupled with lack of clear morphologic or molecular definition.

Since each individual interphase chromatin loop represents a single transcription unit, it was further proposed that in anchoring loops, the scaffold establishes boundaries between chromatin domains. Specific DNA sequences were found to bind the interphase chromosome scaffold *in vitro.* These sequences, termed SARs (an acronym for scaffold attached regions) or MARs (an acronym for matrix attached regions), might therefore be considered to be chromatin boundary elements in DNA. To date, results from *in vivo* tests of this notion have been inconclusive. Nevertheless, the idea that S/MARs represent portions of chromosomes that mediate biologically significant interactions between DNA and proteinaceous structural elements of the interphase nucleus remains an attractive and useful working hypothesis.

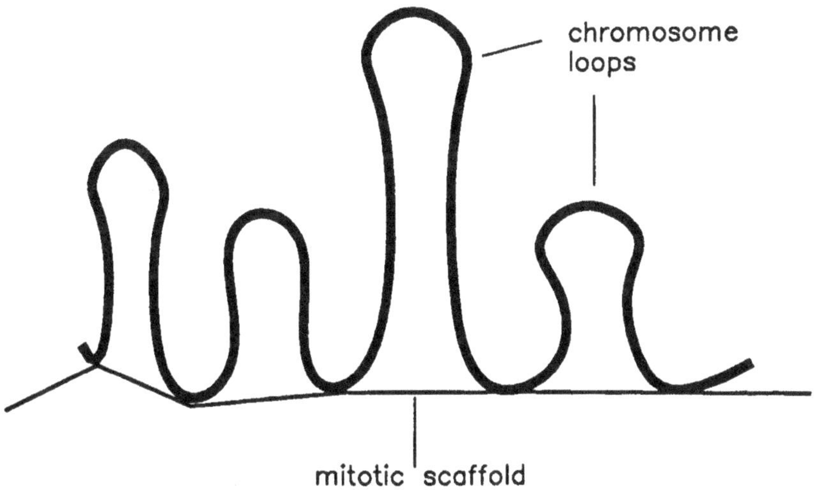

Figure 4. The looped-domain model of mitotic chromosome structure (adapted from Alberts et al., 1989).

Forming a Chromosome

Apart from suggesting that chromosomal loops are anchored to some sort of proteinaceous scaffold or matrix, the looped domain model of higher-order chromatin structure provides relatively little insight into the higher-order structure of the eukaryotic chromosome. Indeed, few other insights have been forthcoming from direct experimentation but models have been proposed to guide further study. One such model, compatible with current notions of chromatin folding during both interphase and cell division (Manuelidis and Chen, 1990; Manuelidis, 1990), postulates that the 30 nm fiber condenses to form a 240 nm (diameter) radial subunit (Figure 5A). Each radial subunit is thought to contain about 30 kilobase pairs of DNA. Two such subunits would contain about 60 kilobase pairs of DNA, similar to the average looped-domain size. It was further postulated that 10 such subunits are arranged circularly to form a 700 nm condensed coil (Figure 5B). This coil would have a hollow center, providing space for scaffold (matrix) components, as well as other proteins (RNA polymerases, DNA polymerases) involved in DNA metabolism.

Manuelidis' model (Figure 5) provides the degree of condensation necessary to account for packing of DNA during both cell division and

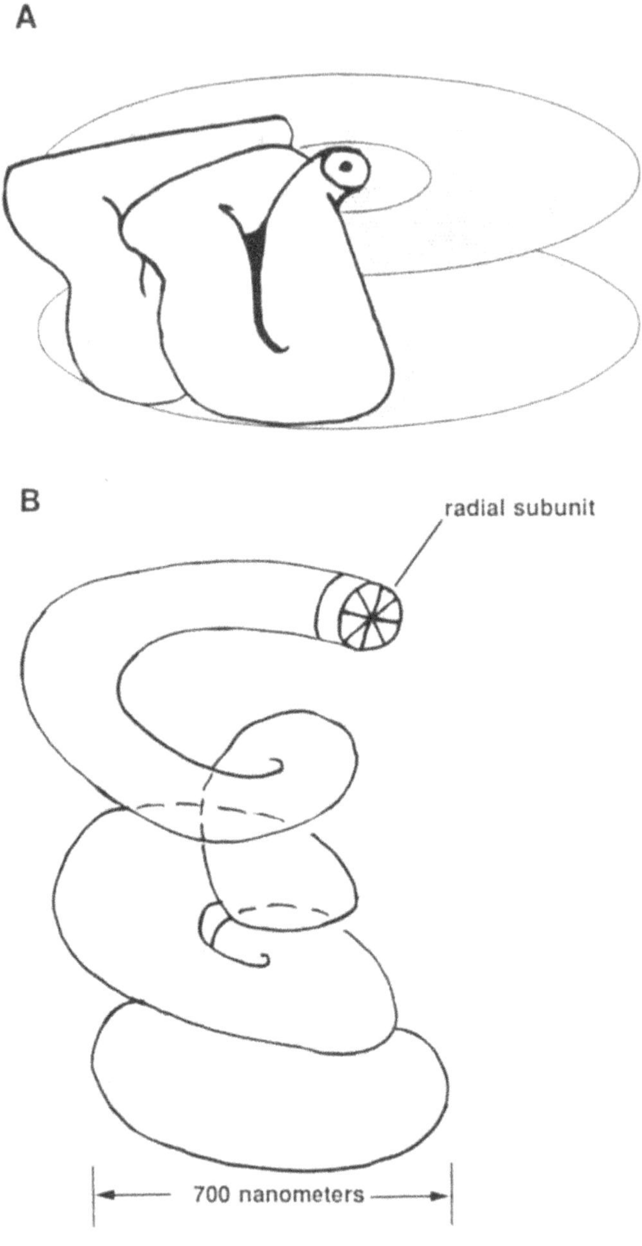

Figure 5. How does chromatin fold to form chromosomes? A model (adapted from Manuelidis and Chen, 1990) showing A, 240-nanometer radial subunits and B, a 700-nanometer coil.

interphase. Moreover, the switch from one state of condensation to another would be mechanistically straightforward. However, it should be stressed that experimental evidence for this model is extremely limited and like the looped-domain model, it represents a potentially useful working hypothesis rather than a final view of the way things really exist.

DNA FOLDING AND THE REGULATION OF TRANSCRIPTION

Nucleosomes and Transcription

Why are we concerned with the higher-order structure of chromatin in the nucleus? Clearly, this question has many answers. However, as articulated earlier, the focus of this chapter is on the role that higher-order chromatin structure plays in the precise and explicit regulation of gene expression. In this regard, histones have long been viewed as general repressors of transcription. Recent results suggest that their role may be much more specific with distinct effects on transcription initiation as opposed to elongation (Kornberg and Lorch, 1992).

Nucleosomes have a profound inhibitory effect on transcription initiation. Although there is apparently considerable variability, i.e. from one gene to another and from one organism to another, in general, histone-containing nucleosomes apparently mask specific regulatory DNA sequence elements with respect to the binding of transcription factors. It is the binding of these transcription factors that controls transcription initiation. Eukaryotic organisms have evolved several mechanisms for overcoming this inhibition of transcription initiation. These include selective positioning of key regulatory DNA sequence elements in regions more likely to be found in linker rather than nucleosomal DNA, and posttranslational modification (e.g., acetylation) of histones that facilitates interactions with selected transcription factors (Lee et al., 1993). In addition, during DNA replication, nucleosomes are displaced from all DNA as part of the synthetic process. Transcription factors can bind to newly replicated DNA and in so doing both block nucleosome formation at specific sites and act much later in time as specific transcription initiators. Currently, this is an area of extremely active research and new details as well as general principles are emerging continuously (Kornberg and Lorch, 1992; Felsenfeld, 1992; Hayes and Wolffe, 1992; Workman and Buchman, 1993).

At first approximation, transcription elongation apparently contrasts with initiation in that actively transcribed genes maintain an overall nucleosomal structure. However, RNA polymerase-containing transcription complexes cannot physically pass through an intact nucleosome. During transcription elongation, there must, therefore, be some disruption of the normal nucleosomal structure of chromatin. At least one strand of the DNA duplex, the so-called coding strand, must dissociate from the nucleosomal core to allow the physical interaction with RNA polymerase inherent in transcription. The exact mechanism or mechanisms whereby such displacement naturally occurs are unknown. At one extreme, it is possible the transcription complex induces only partial disassembly of the nucleosome, and in so doing, gains access to the coding strand of the DNA (Van Holde et al., 1992). Alternatively, it is possible that the nucleosome disassembles completely to facilitate passage of the transcription complex (Kornberg and Lorch, 1991; Adams and Workman, 1993). An attractive model recently put forth (Clark and Felsenfeld, 1992; Morse, 1992) is based on the observation that transcribed DNA is apparently swiveled through the transcription complex rather than that the complex actually moves along the DNA (Liu and Wang, 1987). Accordingly, it was proposed that torsional stress on the DNA generated during transcription provides the driving force for nucleosome displacement (disassembly) in front of the transcription complex.

Higher-Order Chromatin Structure and Transcription

Recent results suggest that molecular details of chromatin structure at the nucleosomal and supranucleosomal level have profound implications for understanding the regulation of transcription. Clearly, higher-order chromatin structure will also be important. However, the paucity of experimental evidence regarding the explicit nature of higher-order chromatin structure makes it particularly difficult to elucidate roles for higher-order structure in transcriptional regulation. This is, nevertheless, an area of intense current interest.

DNA sequences (S/MARs) which appear to interact specifically with elements of the interphase chromosome scaffold also appear to contain or to be coextensive with specific transcription regulatory elements (Cockerill and Garrard, 1986; Gasser and Laemmli, 1986). This has led to the suggestion that transcriptional regulation is related somehow to interaction of chromosomal DNA with the scaffold. To date, details of this putative interrelation have remained obscure. Results of genetic experiments, particularly those obtained in higher eukaryotes such as *Drosophila* (Paro, 1990; Reuter and Spierer, 1992), may ultimately shed considerable light on the role in mammals of higher-order chromatin structure in transcriptional regulation, and consequently, differentiation and development.

SUMMARY AND PERSPECTIVES

In eukaryotes, DNA is folded in a complex manner involving a number of successive stages or levels. Specifics of this folding are highly dynamic varying among cell types and organisms and within a single cell—for example, based on the stage of the cell cycle or local transcriptional activity. Transmission electron microscopy permits distinction between heterochromatin and euchromatin. Heterochromatin is relatively inactive transcriptionally, whereas euchromatin is relatively active. The first level of DNA folding into nucleosome-containing 10 nm filaments is the best understood. As a result, the relationship between nucleosomal organization of chromatin and transcriptional regulation has been at least partially elucidated. Similarly, folding of nucleosome-containing filaments into 30 nm fibers is relatively well-understood. Since packing of chromatin into 30 nm fibers leads to general transcriptional inactivation, the role in transcriptional regulation of this level of higher-order chromatin structure seems clear. Much less clear is the role in transcriptional regulation of chromatin structure above the level of the 30 nanometer fiber. It is apparent that greater elucidation of this question will require detailed molecular characterization of nuclear architecture. To date, such details have not been forthcoming. However, recent technical advances and the availability of a number of specific probes should lead to novel insights in the immediate future.

ACKNOWLEDGMENTS

N.S. was supported by a long-term research fellowship from the International Human Frontier Science Program Organization. P.F. was supported in part by research grants from the National Institutes of Health. It is a pleasure to acknowledge A. Daraio for help in preparation of the manuscript and computer design of figures. This chapter is dedicated to Naomi R. and David W. Fisher on the occasion of their retirement as well as their forty-fourth wedding anniversary.

REFERENCES

Adams, C. C. & Workman, J. L. (1993). Nucleosome displacement in transcription. Cell 72, 305-308.

Alberts, B., Bray, D., Lewis, J., Raff, M., Roberts, K., & Watson, J. D. (1994). Molecular Biology of the Cell. Garland Publishing Inc., New York.

Clark, D. J. & Felsenfeld, G. (1992). A nucleosome core is transferred out of the path of a transcribing polymerase. Cell 71, 11-22.

Cockerill, P. N. & Garrard, W. T. (1986). Chromosomal loop anchorage of the kappa immunoglobulin gene occurs next to the enhancer in a region containing topoisomerase II sites. Cell 44, 273-282.

Felsenfeld, G. (1992). Chromatin as an essential part of the transcriptional mechanism. Nature 355, 219-224.

Felsenfeld, G. & McGhee, J. D. (1986). Structure of the 30 nm chromatin fiber. Cell 44, 375-377.

Fisher, P. A. (1989). Chromosomes and chromatin structure: The extrachromosomal karyoskeleton. Curr. Opin. Cell Biology 1, 447-453.

Gasser, S. M. & Laemmli, U. K. (1986). Cohabitation of scaffold binding regions with upstream/enhancer elements of three developmentally regulated genes of *D. melanogaster.* Cell 46, 521-530.

Gasser, S. M. & Laemmli, U. K. (1987). A glimpse at chromosomal order. Trends Genetics 3, 16-22.

Goldman, M. A. (1988). The chromatin domain as a unit of gene regulation. Bioessays 9, 50-55.

Grunstein, M. (1992). Histones as regulators of genes. Sci. Amer. 267, 68-74B.

Hayes, J. J. & Wolffe, A. P. (1992). The interaction of transcription factors with nucleosomal DNA. Bioessays 14, 597-603.

Jeon, K. W. & Friedlander, M. (1987). Chromosome structure: Euchromatin and heterochromatin. Intl. Rev. Cytology 108, 1-60.

Kornberg, R. D. & Lorch, Y. (1991). Irresistible force meets immovable object—Transcription and the nucleosome. Cell 67, 833-836.

Kornberg, R. D. & Lorch, Y. (1992). Chromatin structure and transcription. Ann. Rev. Cell. Biol. 8, 563-587.

Lee, D. Y., Hayes, J. J., Pruss, D. & Wolffe, A. P. (1993). A positive role for histone acetylation in transcription factor access to nucleosomal DNA. Cell 72, 73-84.

Liu, L. F. & Wang, J. C. (1987). Supercoiling of the DNA template during transcription. Proc. Natl. Acad. Sci. USA 84, 7024-7027.

Manuelidis, L. (1990). A view of interphase chromosomes. Science 250, 1533-1540.

Manuelidis, L. & Chen, T. L. (1990). A unified model of eukaryotic chromosomes. Cytometry 11, 8-25.

Morse, R. H. (1992). Transcribed chromatin. Trends Biochem. Sci. 17, 23-26.

Paro, R. (1990). Imprinting a determined state into the chromatin of Drosophila. Trends Genetics 6, 416-421.

Reuter, G. & Spierer, P. (1992). Position-effect variegation and chromatin proteins. Bioessays 14, 605-612.

van Driel, R., Humbel, B., & De Jong, L. (1991). The nucleus: a black box being opened. J. Cell. Biochem. 47, 311-316.

van Holde, K. E., Lohr, D. E., & Robert, C. (1992). What happens to nucleosomes during transcription? J. Biol. Chem. 267, 2837-2840.

Watson, J. D. & Crick, F. H. C. (1953). Molecular structure of nucleic acids. Nature 171, 737-738.

Wolffe, A. P. (1992). Chromatin: Structure and Function. Academic Press, New York.

Workman, J. L. & Buchman, A. R. (1993). Multiple functions of nucleosomes and regulatory factors in transcription. Trends Biochem. Sci. 18, 90-95.

RECOMMENDED READINGS

Alberts, B., Bray, D., Lewis, J., Raff, M., Roberts, K., & Watson, J. D. (1994). Molecular Biology of the Cell. Garland Publishing Inc., New York.

Grunstein, M. (1992). Histones as regulators of genes. Sci. Amer. 267, 68-74B.

Wolffe, A. P. (1992). Chromatin: Structure and Function. Academic Press, New York.

Chapter 4

The Nucleolus

DANIELLE HERNANDEZ-VERDUN and

HENRIETTE R. JUNERA

Principles of Medical Biology, Volume 2
Cellular Organelles, pages 73-92

ISBN:1-55938-803-X

INTRODUCTION

General Remarks and Definitions

The prominent and refractile structures observed in cell nuclei under phase-contrast microscopy are nucleoli. These structures are the only nuclear organelles distinguishable by light microscopy that can be related to specific genes and functions. The nucleoli are nuclear territories devoted to ribosome biogenesis. They contain specialized units for various steps of ribosomal RNA synthesis and units for assembly of RNA with ribosomal proteins (see p. 81 for details). During evolution, nucleoli appeared concomitantly with nuclear envelope compartmentation. They correspond to the compartmentation of nuclear functions since nucleolar proteins, ribosomal RNAs, and some small RNAs are exclusively found in nucleolar territories. The nucleolus is not a stable organelle. Its structure, size, and organization depend on ribosome biogenesis. The production of ribosomes is directly correlated with protein synthesis and thus nucleolar morphology varies according to cell type, cell differentiation, and the stage of the cell cycle. Various pathologies also affect nucleolar morphology in relation to cell growth and proliferation.

During the cell cycle, nucleoli develop from chromosomal sites, called nucleolar organizer regions (NORs). NORs were first described by B. McClintock in *Zea maize* in 1934 (McClintock, 1934), and have since been shown to contain tandemly arranged ribosomal RNA/ genes (rDNA) and specific NOR-proteins. This chapter will focus on the organization of nucleoli in mammalian cells, and on macromolecular aspects of the nucleolar factory. We review information on where, when, and how the different steps of ribosome biogenesis are organized in nucleoli.

For more detailed information there are several reviews in the literature: for structural organization of the nucleolus (Busch and Smetana, 1970; Goessens, 1984; Hernandez-Verdun, 1986; Sommerville, 1986 Jordan, 1991; Scheer and Benavente, 1990); for nucleolar proteins the most exhaustive recent review is in Olson (1990); for ribosomal gene function (Warner, 1990; Raué and Planta, 1991; Sollner-Webb and Mougey, 1991); and finally the book by Hadjiolov is an exhaustive overview of the subject (Hadjiolov, 1985).

ORGANIZATION OF THE NUCLEOLUS IN HUMAN CELLS

Number and Size

Mammalian nuclei contain one or a few nucleoli. The number of nucleoli in HeLa cells in culture is between 1 and 6 with considerable size variability between them. However, the number of nucleoli cannot be accurately determined since nucleoli fuse in human cells shortly after entry into interphase. This phenomenon has been illustrated by video microscopy during other interphase periods. Nucleoli vary in form, staining-capacity, and behavior, thus rendering their classification difficult by light microscopy alone. The shape of nucleoli can differ markedly from one cell type to another, and even within a single cell. They are generally stained by acidic dyes for histology but this staining capacity can vary: for example, during germ cell differentiation nucleoli can be stained with basic dyes. In light microscopy, nucleoli include small deeply stained granules or nucleolini. In human cells, the size of nucleoli varies from less than 0.5 μm in diameter in mature lymphocytes to 3-9 μm in proliferating cells.

Polarity

Nucleoli are heavily involved in nuclear polarity. They are traditional cytological landmarks, and there is also evidence for polarity between different nucleoli, between nucleoli and the nuclear envelope, and between nucleoli and centromeres.

In the metaphasic plate, the human NOR-bearing chromosomes are closer to each other than would be expected if the chromosomal distribution were random. This observation indicates that there is certainly a polarized arrangement of the NOR-bearing chromosomes in interphasic nuclei. Clearly, the NOR-association is consistent with nucleolar fusion which occurs in species that possess several pairs of NORs (see p. 87). After fusion or in species without nucleolus-fusion, the position of the nucleoli remains fairly stable, even in rotating nuclei.

Very frequently, nucleoli in higher eukaryotic cells are located at or near the nuclear envelope. This position seems to be related to the presence of specific skeletal structures at the site of nucleolar attachment to the envelope. This nucleolar skeleton either adheres directly to the nuclear lamina or is attached to it by a pedicle as visualized in spread lamina preparations.

In nuclei with centrally located nucleoli, the nuclear envelope is folded to form the nucleolar canal which is in direct contact with the nucleoli (Figure 1). This canal was shown to be a nuclear envelope specialization that depends on the presence of active NORS. It is possible that the nuclear canal, as well as the close relationship between

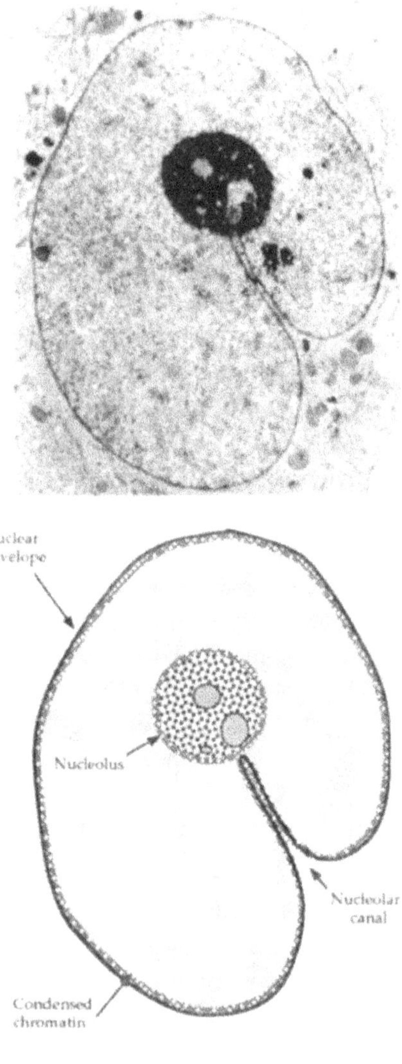

Figure 1. (**a**) Section of HeLa nucleus observed in electron microscopy. (**b**) Schematic presentation of the image seen in **a**. Note the relationship between nuclear envelope and nucleolus.

the nucleoli and the nuclear envelope, constitute nuclear envelope specializations that favor nucleo-cytoplasmic exchanges.

Fine Structure

The fine structure of nucleoli is clearly visualized only by electron microscopic observation of thin sections. The main disadvantage of thin sections is that the three-dimensional arrangement of the different nucleolar territories cannot be seen. Nucleoli are exceedingly robust and withstand high ionic strength, detergents, and sonication. Little is known as yet of the molecules or structures that hold nucleoli together. In transmission electron microscopy, nucleoli appear to be mainly composed of fibrils and granules. The granules are about 150 nm thick and constitute the granular component (GC) of nucleoli. The fibrils display two different electron opacities. Some fibrils are very dense and have been called the dense fibrillar component (DFC); others constitute light areas which have been called fibrillar centers (FC) (see nucleolar nomenclature reviewed in Jordan, 1984). Most often, FCs are surrounded by the DFC which is in close contact with the GC (Figure 2). At the periphery of nucleoli highly contrasted structures are frequently observed: these are condensed chromatin. Condensed chromatin is also found inside nucleoli in the GC and clearly visible close to FC, but not in FC or DFC in mammalian cells. Analysis of nucleolar chromatin by specific DNA staining indicates that decondensed chromatin is present in the FC of nucleoli.

The high electron opacity of the biological structures in conventional transmission electron microscopy is mainly due to staining with uranyl and lead, two heavy metals used to contrast the structures. In nucleoli, the high contrast observed after uranyl and lead staining seems mainly due to nucleic acids since the same high contrast is observed after blockage of the binding sites in proteins.

Types of Nucleolar Organization

The three nucleolar domains, FC, DFC, and GC are found in all mammalian cell nucleoli except those in oocytes, spermatocytes, and cell embryos at the first stages of development. A classification has been established for mammalian nucleoli according to the organization of these components: reticulated, compact, segregated, or ring-shaped nucleoli (Figure 3). The type of nucleolar organization correlates with the activity of the cells with respect to ribosomal biogenesis: compact

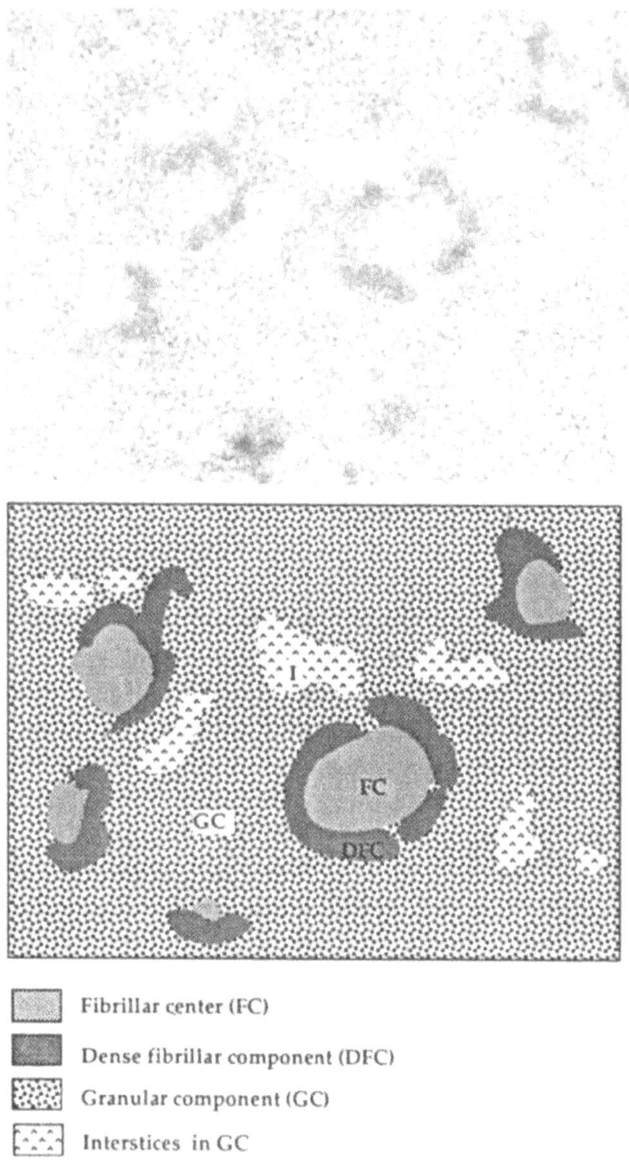

Figure 2. (**a**) Detail of nucleolar structures observed in electron microscopy in a HeLa cell. Magnification 50,000. (**b**) Schematic presentation of the image seen in **a**. Fibrillar center (FC) is surrounded by a dense fibrillar component (DFC) in close contact with granular component (GC).

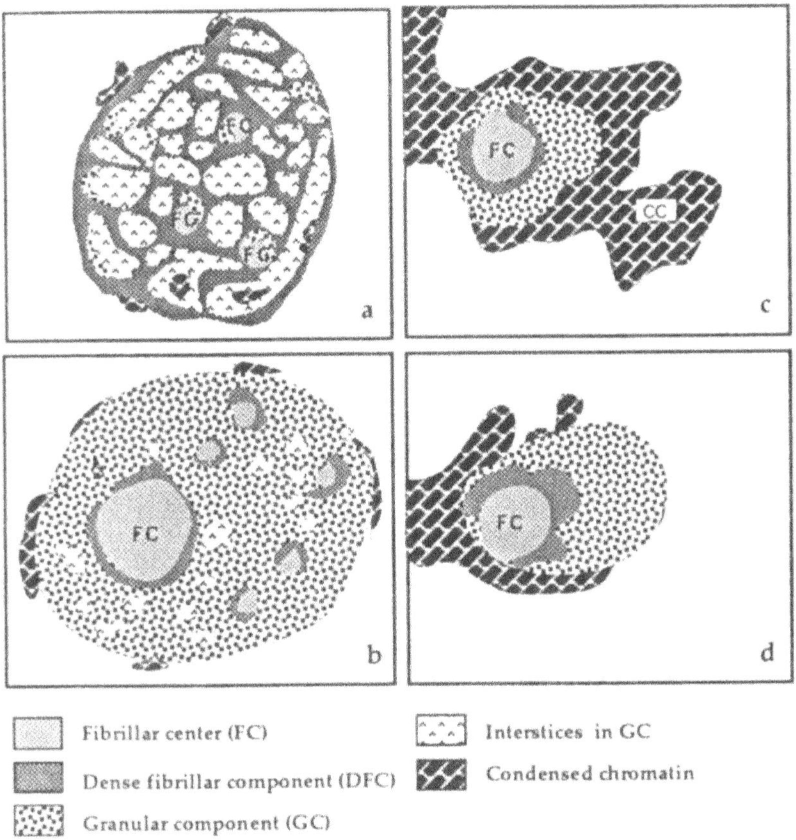

Figure 3. Schematic presentation of 4 types of nucleolar organization. (**a**) Reticulated nucleolus in active cells; (**b**) compact nucleolus in active cells; (**c**) ring-shaped nucleolus in mature lymphocyte; (**d**) segregated nucleolus in drug-induced rRNA synthesis inhibition.

and reticulated nucleoli are found in actively transcribing cells; segregated and ring-shaped nucleoli are found in quiescent cells or cells treated with drugs that inhibit transcription.

1. Reticulated nucleoli are characterized by a network of threads of equal thickness which are made up of either DFC or GC. The network has been called the nucleolonema. The FC are most often small. Reticulated nucleoli are found in very active cells and in cancer cells.

2. Compact nucleoli are characterized by a compact organization of the granular component in which some interstices can be seen; they contain several fibrillar centers surrounded by DFC which does not form a network in the thickness of one section. The sizes of the FC can vary greatly. Compact nucleoli are found in active cells and often in cell lines in culture.
3. Segregated nucleoli are characterized by separation of the nucleolar component which remained superimposed but no longer intermingled. They are found after blocking RNA synthesis under physiological conditions and after various drug treatments.
4. Ring-shaped nucleoli are characterized by one FC surrounded by a small amount of DFC, which in turn is surrounded by granules. They are always small and are found in nontranscribing cells such as in mature lymphocytes.

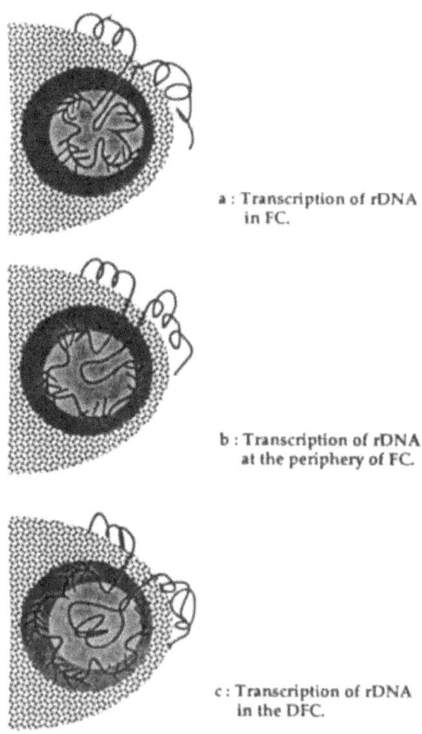

Figure 4. Different hypotheses for the location of the rDNA transcription in nucleoli

Relationship Between Structures and Function in Nucleoli

Traditionally, the FC is considered to be the storage site of nontranscribed ribosomal genes, the DFC the site of the transcription of these genes and the GC, the site of the maturation and storage of the ribosomal subunits. However, the site of transcription is controversial (Jordan, 1991). Some authors believe that transcription occurs in the FC, others, at the border between the FC and DFC and still others, in the DFC only (Figure 4).

MACROMOLECULAR ASSEMBLY OF NUCLEOLI

Ribosomal Genes and Ribosomal RNAs

Ribosomes contain four different ribosomal RNAs: 18S RNA in the small subunits; and 28S, 5.8S, and 5S RNAs in the large subunits. The ribosomal genes (rDNA) corresponding to 18S (1900bp), 5.8S (160bp), and 28S (5100bp) RNAs are localized in the nucleoli. The primary trancripts are 45S ribosomal RNAs which are methylated and cleaved to give 18S, 5.8S and 28S ribosomal RNAs (Figure 5). The 5S (120bp)

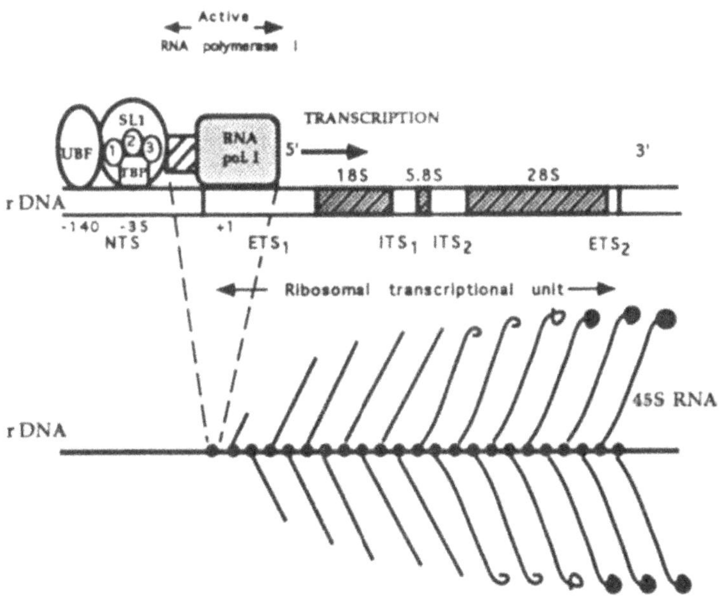

Figure 5. Schematic presentation of ribosomal genes (rDNA) and associated proteins in relationship with transcription. NTS: nontranscribed spacer; ETS external transcribed spacer; ITS: internal transcribed spacer; UBF, SL1, TBP: transcription factors (see text).

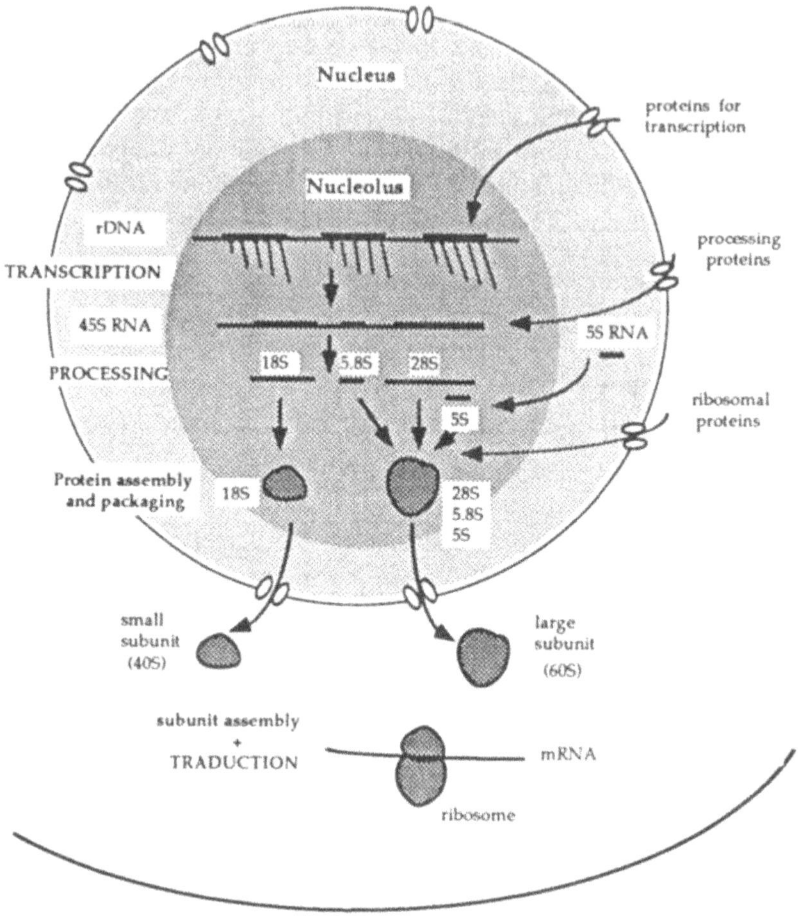

Figure 6. Ribosomal biogenesis pathway (simplified) from the transcription of rDNA to the assembly of ribosomes.

RNAs are synthesized outside of the nucleoli and migrate to the nucleoli before being exported into the cytoplasm (Figure 6).

The ribosomal genes are repetitive sequences. For example, HeLa cells contain 200 copies per haploid genome distributed among five chromosome pairs. However, not all of these copies are expressed simultaneously. The rDNA is transcribed by RNA polymerase I. The initiation of transcription by RNA polymerase I is controlled by ribosomal gene-specific transcription factors; some of them are species specific. For example, mouse ribosomal genes can be transcribed only in the presence of mouse transcription factors which have not yet been completely identified.

Electron microscopy of spread chromatin provides evidence of active transcription of rDNA. The first images were obtained by Miller and Beatty (1969). The active transcription units looked like Christmas trees. The rDNA forms the axis of the transcription units and the nascent rRNAs constitute lateral fibrils of different sizes attached to the axis. Each active transcription unit is separated from the next by an apparently nontranscribed spacer. On the transcribed part of the axis, the granules correspond to RNA polymerase I. The 5' ends of the nascent rRNA fibrils appear as terminal knobs that are probably the cleavage processing complexes (see Figure 5).

The assembly of rRNA with ribosomal proteins takes place in the nucleoli, probably in GC. The preribosomal subunits migrate independently into the cytoplasm; the small subunits containing 18S migrate more rapidly. The small and large subunits are then assembled in the cytoplasm.

Specific small nucleolar ribonucleoproteins (snoRNPs) are found in the nucleoli. They contain four different RNAs (U3, U8, U13, and U14) not found in the nucleoplasmic snRNPs. U3 snoRNAs are precipitated by antifibrillarin antibodies and participate *in vitro* in the first cleavage of 45S RNA at the processing sites. It has been suggested that U3-containing RNP particles are the distinct electron dense terminal knobs observed at the 5' end of the nascent RNAs on molecular spread transcriptional units (see above). In the yeast nucleus, where the number of snRNAs seems to be much larger than in mammals, several specific nucleolar snoRNPs (snR3-snR5, snR8-snR10 and snRl28) were found to associate with various ribosomal RNA precursors. Although the precise roles of these nucleolar snoRNPs in ribosome biogenesis are not yet known, it is clear that their functions differ from those of nucleoplasmic snRNPs, which are involved in the splicing of pre-mRNAs (Lührmann, 1990).

Proteins

A nucleolar-specific protein can be defined as a protein which is specifically located in the nucleoli and is involved in ribosome biogenesis. The nucleolar-specific proteins do not end up in the mature ribosome even though they may be engaged in a shuttle process between nucleoli and the cytoplasm. It is possible to prepare pellets of almost pure isolated nucleoli by sonication of isolated nuclei and subsequent centrifugation in 0.88 M sucrose (Zalta et al., 1971). This procedure has been used to characterize various nucleolar proteins and the composition of the different nucleolar territories by sequential extractions. In the seventies, the first paper in this field (Orrick et al., 1973) described 97 nucleolar proteins by two-dimensional gel electrophoresis. As this study only detected major nucleolar proteins below 120 kD, there are presumably several hundred proteins specifically confined within the nucleolar territory. Nucleolar proteins have been suggested as being involved in transcription, maturation, packaging, and transport of ribosomal particles (Reeder, 1990); but currently the role of the great majority is only conjectural. The nuleolar proteins which have been characterized in some detail are RNA polymerase I, UBF, SLl complex, nucleolin, B23, and fibrillarin (Table 1).

RNA polymerase I is the enzyme that specifically transcribes the ribosomal gene for the large ribosomal RNAs. This polymerase is a large molecule (400 to 700 kD) with complex subunit structures and is composed of at least 6 polypeptides (Sentenac, 1985). Its two large

Table 1. Distribution of Nucleolar Proteins (Simplified)

		Localization		
Proteins	*MW (kD)*	*Interphase*	*Mitosis*	*Role and remarks*
RNA pol I	180-120-60 42-29-25	FC	NOR	Enzyme of transcription
UBF	97-94	DFC+FC	NOR	transcription factor
Ag-NOR	X	DFC+FC	NOR	*transcrip/ prolifer marker
Ki-67	345-390	DFC	PC	proliferation marker
Fibrillarin	34	DFC+GC	PC	processing/ complex with U3
B23	38	DFC+FC	PC	*matur/assembly/ shuttle protein
Si	45	GC+cyt	PR	ribosomal protein
Nucleolin	100	DFC+GC	cyt	*transcrip/ RNA packaging shuttle protein

Notes: Abreviations: RNA pol I: RNA polymerase I; UBF: upstream binding factor; Ag-NOR proteins: a group of silver stained nucleolar proteins under investigation; Sl: ribosomal protein; PC: periphery of the chromosomes; PR: perichromosomal region; cyt: cytoplasm; transcrip: transcription; prolifer: proliferation; matur: maturation; *putative role.

components, which are highly conserved between yeast and mammals, are involved in the polymerization reaction and have DNA binding properties in yeast. The small subunits seem to have accessory roles including the binding to transcriptional factors, nuclear localization or enzyme assembly. In mammalian cells, the two large subunits are between 180-197 kD and 120-130 kD depending on the species. The small subunits are around 60 kD, 42 kD, 29kD, and 25 kD. RNA polymerase I activity is increased in more rapidly growing mammalian cells and is proportional to growth rates in tumor cells. There is an equilibrium between the free and bound fractions of RNA polymerase I that is maintained upon stimulation of cell growth rate, or after hormonal or drug treatment (Sentenac, 1985).

Active RNA polymerase I is required for the formation of the nucleolus as it is one of the major components but RNA polymerase I is not able to bind to rDNA in the absence of transcription factors. The initiation of RNA polymerase I transcription is dependent on the presence of nucleolar transcription factors including UBF (upstream binding factor) (Jantzen et al., 1990) and other factors which have been given different names (SL1 or TIF-IB complex, TIF-IA, TIF-C). UBF and TIF-IB interact with the rDNA promoter. Purified human UBF is composed of two closely related polypeptides (94-97 kD). UBF appears to be present at approximately 50,000 copies per cell which is very high when compared to the approximately 200 copies of rDNA per haploid genome in HeLa cells. The transcription of ribosomal genes is regulated either by a modification of RNA polymerase I which prevents its interaction with transcription factors and/or by the action of activating or repressing transcription factors. This provides a versatile mechanism for adaptation of rDNA transcription to cell needs. In the nucleoli, immunodetection of RNA polymerase I and UBF visualizes small positive beads that could correspond to individual transcriptional units. In electron microscopy, anti-RNA polymerase I antibodies bind in FC, whereas UBF antibodies are observed in DFC and FC. At the end of G2 phase, the transcription switch off is concomitant with the gathering of UBF and RNA polymerase I at mitotic NORs.

Nucleolin, also called C23 (Mr = 100-110 kD; pI = 5.2), is a major nucleolar protein in the nucleoli with active ribosomal biogenesis (10% of total nucleolar protein in CHO cells). This protein is found in all eukaryotes and its phosphorylation varies both through the cell cycle and with cell growth rates. The synthesis of nucleolin is generally induced during cell proliferation. The maturation of nucleolin in the nucleoli, presumably by a proteolytic activity into a 50 kD stable polypeptide appears to be linked to the nucleolin function. Nucleolin comprises three distinct domains: an acidic amino-terminal region, four RNA-binding domains and an arginine-glycine-rich (RG-domain) carboxyl-terminal end. Nucleolin is thought to participate in the early processes of ribosome biogenesis, such as the modulation of the chromatin conformation in the nucleolus and binding to nascent rRNA, but its function is not entirely clear. Nucleolin is found in DFC and GC. It was recently shown to shuttle between the nucleoli and the cytoplasm. Thus this protein seems able to follow the sequence of events ensuring ribosome biogenesis, but is not part of the final product.

The nucleolar protein B23, also called numatrin or No38 (Mr = 38 kD; pI = 5.1), is a major RNA-associated phosphoprotein which is believed to be one of the factors responsible for preribosomal particle assembly. Increased rates of synthesis of protein B23 are correlated with cell proliferation. This protein is a member of the nucleoplasmin family, and is widely distributed in higher eukaryotes: in all cases it has an apparent molecular weight of 37-38 kD. It binds cooperatively with high affinity to single-stranded nucleic acids and exhibits RNA helix destabilizing activity. Protein B23 is mostly located in the GC and is associated with the most mature nucleolar preribosomal RNP. It was found to migrate out of the nucleoli when RNA synthesis decreases during serum starvation. This protein, which appears to be involved in the intermediate stages of ribosome assembly, also seems, like nucleolin, to shuttle between the nucleoli and cytoplasm.

Fibrillarin, also called NOP1 in yeast (Mr = 34-36 kD; pI = 8.5), is a nucleolar protein located in the DFC. Fibrillarin has been conserved very similarly from yeast to human. It is essential for cell growth, since the cells are not viable in the absence of the corresponding gene. It became a very fashionable nucleolar protein when it was demonstrated that it associates with U3 small nucleolar RNA and subsequently with the U8 and U13 snRNAs (Tollervey and Hurt, 1990).

Shuttle Nucleolar Proteins and Nucleolar Targeting Signal

Various nuclear proteins have been shown to shuttle between the nucleus and cytoplasm. These proteins seem to be involved in import and/or export from the nucleus. Three nucleolar proteins have been proposed to be shuttle proteins: nucleolin, B23, and Noppl40. Noppl40 is a nucleolar phosphoprotein of 140 kD (Meier and Blobel, 1992) that has been identified and purified as a protein which binds with the nuclear localization signal (NLS) sequences. The NLS is able to address molecules to the nuclei. It has been suggested that Noppl40 functions as a chaperone for import into and/or export from the nucleolus (Meier and Blobel, 1992).

The localization of the nucleolar proteins in specific regions of the nucleus indicates that these proteins must contain signals that determine their final destination in addition to NLS sequences. Amino acid sequences that target proteins to the nucleolus, i.e. nucleolar targeting signals (NOS), have been described in viral proteins including the HIV-1 tat and rev proteins. For some nucleolar proteins, distinct nucleolar localization signals have been described in addition to an NLS: they consist of a series of arginine and glycine residues (RG-domain). These domains are found in nucleolin and fibrillarin, but not in Noppl40. The RG-domain seems to be involved in RNA recognition rather than being a true signal. For nucleolin both the RG-domain and the RNA- binding sequences are needed to drive the protein to nucleoli. These data support the hypothesis that the product (most probably the RNA) organizes the factory. However, this field of research is only beginning and needs further investigation.

NORS

Mitotic NORs

In mitotic chromosomes, NORs are characterized by the presence of rDNA in association with specific proteins. rDNA in human genome is distributed among five NOR-bearing chromosome pairs (i.e., 13, 14, 15, 21, and 22). The NORs correspond to the secondary constrictions of the chromosomes in human (Figure 7) and most other species. The number of NORs varies between species. For example, in PTK_1 there is only one NOR-bearing chromosome pair, the X-chromosome pair. Consequently, males carry only one NOR and it is sufficient to ensure the synthesis of 45S RNAs and assembly of ribosomal subunits for the cells. In cells that contain multiple NORS, the NORs do not all have equivalent transcriptional activity. Active and inactive NORs can coexist in the same cell; the inactive NORs lack some proteins found in active NORS.

Figure 7. NOR-bearing chromosomes in human karyotype.

What Are NOR Proteins?

Proteins associated with NORs have been detected either by Ag-NOR staining or using antibodies. Characterization of the NOR proteins remains incomplete. The easiest way to locate the NORs on chromosomes is by Ag-NOR staining. This technique is based on silver reduction under acidic conditions and does not stain most other cellular proteins. However, various NOR-specific proteins, called Ag-NOR proteins are revealed (Goodpasture and Bloom, 1975). The technique is specific, rapid, and quantitative, and consequently is widely used. Because these proteins are required for ribosomal gene transcription, this staining is used to identify NORS. It has been demonstrated in somatic hybrids that only the chromosomes of the species which express the rDNA during interphase possess Ag-NOR proteins. Ag-NOR proteins are thus markers of active NORS. There is currently no satisfactory way of identifying Ag-NOR proteins on mitotic chromosomes. It has been proposed that nucleolin and protein B23 are the major AG-NOR proteins during interphase (for review see Olson, 1990). This cannot be the case during mitosis because it is known that these proteins are not associated with the NORs during mitosis (Table 1). In addition, we know that not all NOR proteins are revealed by Ag-NOR staining.

Antibodies directed against RNA polymerase I and UBF have been used to demonstrate that these proteins remain associated with the NORs during mitosis. Thus, it seems more likely that proteins directly involved in the transcriptional machinery remain bound or associated to rDNA during mitosis. However, the presence of proteins in association with ribosomal genes does not mean that they are active or linked to the genes. Nevertheless, the sequestration of nucleolar proteins of the transcription machinery, UBF, RNA polymerase I, close to the genes during mitosis could allow the rapid switch-on of ribosomal transcription in telophase.

CELL CYCLE

Nucleoli are dismantled during cell division, and are reassembled in telophase. During the cell cycle, rDNA transcription starts in telophase and stops in late G_2 phase when the chromatin condenses into chromosomes. The entry into the mitotic phase (M phase) coincides with the specific phosphorylation of major nucleolar proteins, including nucleolin and B23, by $p34^{cdc2}$ kinase. It has been suggested that this phosphorylation controls mitotic changes in nucleolar protein localization and activity. If true, this would mean that nucleolar dispersion is controlled by the same mechanisms as those that control chromosome condensation, spindle formation, and nuclear envelope breakdown.

Relocalization of Nucleolar Proteins During Mitosis

In prophase nucleoli disappear, often after becoming smaller or undergoing fragmentation. In some cases they are ejected into the cytoplasm while the nuclear membrane disappears. These structures, called residual nucleoli, have been observed in cells with very active ribosome biogenesis, for example, Chinese hamster ovary cells.

When nucleoli are dismantled, their proteins become differently distributed within the dividing cells. Some proteins remain associated in NORS, some disperse to the peripheries of all chromosomes, while others are scattered throughout the cytoplasm without any detectable sites of accumulation (Sommerville, 1986). During telophase these disparate nucleolar constituents rapidly reassemble in the NORs, thereby reforming functioning nucleoli.

Nucleolar Protein Markers of the Cell Cycle

Some nucleolar proteins appear to be cell cycle-dependent because they are only found in cycling cells or because they are regulated by the cell cycle. The proteins in the first category can be considered to act as nucleolar markers of the cell cycle by their presence or their level of accumulation in the nucleolus.

The first nucleolar cell-cycle marker was identified using the Ki-67 monoclonal mouse antibody. This monoclonal antibody is directed against nucleolar proteins of high molecular weight (345/395 kD) whose function is unknown. It is the standard marker for the characterization of cycling cells in pathology.

Several other proliferation-associated nucleolar proteins have been described as 145kD and 125kD proteins, the 86-70 kD complex, and p120 and p40 proteins. However, the function of these proteins is not yet clear since injections of the corresponding antibodies inhibit both DNA and RNA synthesis. Ag-NOR protein concentrations in nucleoli can also be indicators of cell proliferation. In the absence of proliferation in differentiated cells, Ag-NOR protein concentrations are greatly reduced, whereas in proliferating cells they are high. It is unknown whether this increase is due to the accumulation of the major Ag-NOR proteins or synthesis of additional Ag-NOR proteins.

NUCLEOLI AND PATHOLOGY

Nucleoli are indicators of cell activity. In fact the first morphological modification in cells observed after inhibition of transcription is the modification of the nucleolar organization. The drug, actinomycin D, inhibits the activity of RNA polymerase I. This inhibition is reflected by the disorganization of nucleoli: they become segregated. Therefore, a segregated nucleolus indicates that rDNA transcription has been stopped. In contrast, high cell activity is associated with large nucleoli. This is evidenced in many cancer cells in which nucleoli are generally bigger than in normal cells. This can be used to recognize cancer cells in a tissue. Furthermore, cancer cell nucleoli also contain high concentrations of nucleolar proteins including Ag-NORs proteins, and the proliferating-nucleolar markers Ki-67 and p120. Recently, it has been reported that Ag-NOR protein levels have a prognostic value for human cancer.

Nucleolar proteins are the targets for autoimmune antibodies (Tan, 1989). Most frequently, the patients suffer from scleroderma and rheumatoid diseases. All three nucleolar components can be involved and the antibodies cross-react with antigens conserved in other species sometimes as unrelated to human as yeast nucleolar proteins. Autoimmune sera directed against nucleoli contain antibodies which recognize RNA polymerase I, UBF, nucleolin, fibrillarin or protein B23, as well as many other nucleolar proteins. Autoantibodies against fibrillarin have been found in human patients with scleroderma and can also be induced in mice by mercuric chloride treatment.

SUMMARY

Nucleoli are nuclear territories devoted to ribosome biogenesis. Their shape, size and organization vary with cell activity. They are larger in cancer cells than in normal cells. In electron microscopy, they exhibit three main components: FC, DFC and GC, corresponding to different steps of ribosome biogenesis. Active nucleoli are either reticulated or compact, whereas inactive nucleoli are ring-shaped or segregated.

The ribosomal genes, rDNA, are localized in NORs of different chromosomes. The number of NORs depends on the species. There are 5 pairs in humans, each with different activities. rDNA is transcribed by RNA polymerase I in association with transcription factors, including UBF and the SLl complex.

In nucleoli, there is transcription of 45S rRNAs, processing of 18S, 5.8S, and 28S rRNA, transport of 5S rRNA and assembly of ribosomal proteins and RNAs into preribosomal subunits. The processing involves U3 snoRNAs in a processing complex which contains fibrillarin.

The major nucleolar proteins in active nucleoli are nucleolin and protein B23. Autoantibodies have been found directed against each of the characterized nucleolar proteins.

REFERENCES

Busch, H. & Smetana, K. (1970). The Nucleolus. Academic Press, New York.

Goessens, G. (1984). Nucleolar structure. Intl. Rev. Cytol. 87, 107-158.

Goodpasture, C. & Bloom, S. E. (1975). Visualization of nucleolar organizer regions in mammalian chromosomes using silver staining. Chromosoma 53, 37- 50.

Hadjiolov, A. A. (1985). The nucleolus and the ribosome biogenesis. In: C.B. Monographs, vol. 12, pp. 1-268. Springer-Verlag, New York.

Hernandez-Verdun, D. (1986). Structural organization of the nucleolus in mammalian cells. In: Methods and Achievements in Experimental Pathology, 12 (Jasmin, G. & Simard, R., eds.), pp. 26-62. Karger, Basel.

Jantzen, H.-M., Admon, A., Bell, S. P., & Tjian, R. (1990). Nucleolar transcription factor hUBF contains a DNA-binding motif with homology to HMG proteins. Nature 344, 830-836.

Jordan, E. G. (1984). Nucleolar nomenclature. J. Cell Sci. 67, 217-220.

Jordan, E. G. (1991). Interpreting nucleolar structure. Where are the transcribing genes. J. Cell Sci. 98, 437-449.

Lührmann, R. (1990). Functions of U-snRNPs. Mol. Biol. Rep. 14, 183- 192.

McClintock, B. (1934). The relation of particular chromosomal element to the development of the nucleon in Zea maize. Z. Zelforsch. mikrosk. Anat. 21, 294-328,

Meier, U. T. & Blobel, G. (1992). Noppl40 shuttles on tracks between nucleolus and cytoplasm. Cell 70, 127-138.

Miller, O. L. & Beatty, B. R. (1969). Visualization of nucleolar genes. Science 164, 955-957.

Olson, M. O. J. (1990). The role of proteins in nucleolar structure and function. In: The Eukaryotic Nucleus. Molecular biochemistry and macromolecular assemblies, 2 (P.R. Strauss & S.H. Wilson ed.,), pp. 519-559. The Telford Press. Caldwell, New Jersey.

Orrick, L. R., Olson, M. O. J., & Busch, H. (1973). Comparison of nucleolar proteins of normal rat liver and Novikoff hepatoma ascites cells by two-dimensional polyacrylamide gel electrophoresis. Proc. Nat. Acad. Sci., USA 70, 1316-1320.

Raué, H. A. & Planta, R. J. (1991). Ribosome biogenesis in yeast. Prog. Nucleic Acid. Res. 41, 89-129.

Reeder, R. H. (1990). rRNA synthesis in the nucleolus. Trends Genet. 6, 390-395.

Scheer, U. & Benavente, R. (1990). Functional and dynamic aspects of the mammalian nucleolus. BioEssays 12, 14-21,

Sentenac, A. (1985). Eukaryotic RNA polymerases. CRC Crit. Rev. Biochem. 18, 31-90.

Sollner-Webb, B. & Mougey, E. B. (1991). News from the nucleolus: rRNA gene expression. 16, 58-62.

Sommerville, J. (1986). Nucleolar structure and ribosome biogenesis. TIBS 11, 438-442.

Tan, E. M. (1989). Antinuclear antibodies: Diagnostic markers for autoimmune diseases and probes for cell biology. Adv. Immun. 44, 93-151.

Tollervey, D. & Hurt, E. C. (1990). The role of small nucleolar ribonucleoproteins in ribosome synthesis. Molec. Biol. Rep. 14, 103-106.

Warner, J. R. (1990). The nucleolus and ribosome formation. Curr. Opin. Cell Biol. 2, 521-527.

Zalta, J., Zalta, J. P. & Simard, R. (1971). Isolation of nucleoli. A method that combines high yield, structural integrity, and biochemical preservation. J. Cell Biol. 51, 563-568.

Chapter 5

Centromeres and Telomeres

J.B. RATTNER

CENTROMERES

Terminology

Centromere

A constricted portion of the chromosome which is the point of attachment of the spindle microtubules. The centromere divides the chromosome into two portions and its position is constant for a specific

Principles of Medical Biology, Volume 2
Cellular Organelles, pages 93-120

ISBN:1-55938-803-X

chromosome. One type of chromosome classification is based on position of the centromere: near one end, acrocentric, near the center, metacentric, in between is termed submetacentric.

Primary Constriction

A cytological term synonymous with centromere.

Kinetochore

That portion of the centromere that functions specifically to mediate the interaction between the chromosome and the spindle microtubules. In human chromosomes this is a trilaminar structure found on the outer surface of the centromere.

Heterochromatin

Chromatin that remains tightly condensed throughout the cell cycle. The chromatin of the centromere is often referred to as centromeric heterochromatin which denotes its highly condensed nature and its general lack of transcriptional activity.

Repetitive DNA

Nucleotide sequences that are present in the genome in numerous copies.

Satellite DNA

DNA which appears as a discrete band when DNA is centrifuged to isopycnic equilibrium on a cesium chloride density gradient. This DNA is highly repetitive, mostly found in heterochromatin and is generally not transcribed. Many satellite DNA sequences localize to the centromere.

Centromere Organization Composition and Function

The Chromosome is a Linear Structure

In most eukaryotes, including humans, the chromosomal organization of the genome becomes visible at cell division. During this period each chromosome consists of two chromatids. Each chromatid is in turn composed of a single strand of DNA that is folded in a specific and reproducible manner. The ends of each chromatid are known as the **telomere** and the **centromere**, which appears as a constricted region in the chromatid, divides each chromatid into two arms (Figure 1). Short arms have been given the designation **p** for petite and long arms have been given the designation **q**. The centromere is a multifunctional chromosomal region responsible for modulating sister chromatid interaction, ensuring chromosomal integrity, integrating the chromosome with spindle microtubules, and mediating chromosome movement. Fusion of chromosomes or duplication of chromosome sequences can lead to chromosomes with multiple centromeres. These chromosomes are referred to as **multicentric**. When only one of the centromeres function in spindle association, the multicentric chromosome is stable and can be seen to have multiple constrictions at metaphase. If more than one centromere functions in spindle attachment, each chromatid can associate with two poles simultaneously. At anaphase this leads to stretching and breakage of the chromatid.

The Chromosome is the Result of Many Levels of Chromatin Packaging

The path of the DNA fiber within each chromatid is now known at least in terms of its basic arrangement. In eukaryotic cells, double-stranded DNA is packaged with histone proteins to form a basic 10 nm nucleosomal chromatin fiber that has a "beads on a string" appearance. This fiber is in turn folded to form a 30 nm fiber (Figure 2A) that represents the basic DNA arrangement found in both metaphase chromosomes and interphase nuclei. Particularly prominent within metaphase chromosomes, tandem arrays of 30 nm fiber loops are condensed to form a compact fiber approximately 250 nm in diameter (Figure 2B). These 30 nm fiber loops are thought to be associated with the nuclear matrix in interphase nuclei and a protein scaffold in metaphase chromosomes. The 250 nm fiber is detected at the onset of condensation associated with the appearance of metaphase chromosomes (prophase) and the coiling of this fiber, in concert with the condensation of successive coils, results in the final form of the arms of each chromatid (Figure 2C). Gene and protein mapping studies suggest that the coils of sister chromatids have opposite handedness (for details see Rattner and Lin, 1985a; Boy de la Tour and Laemmli, 1988; Manuelidis and Chen, 1990).

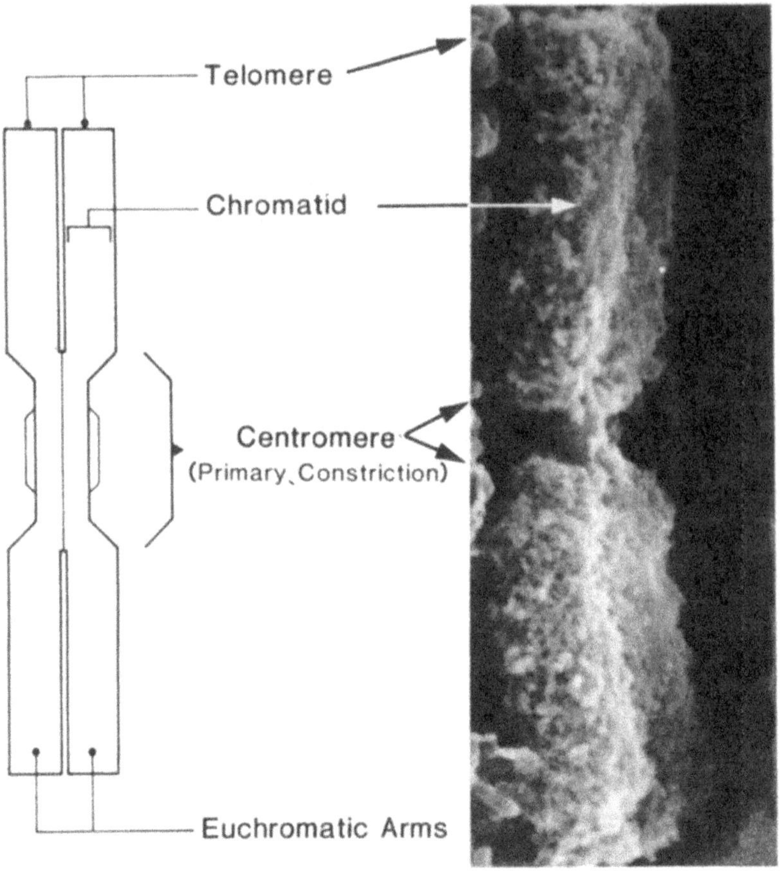

Figure 1. The organization of a metaphase chromosome. **(A)** Diagrammatic representation of a metaphase chromosome denoting its components. **(B)** Scanning electron micrograph of an Indian muntjac metaphase chromosome illustrating the components denoted in Figure 1A. (Figure 1B reproduced from Rattner, 1991).

The Centromere has a Unique Pattern of Chromatin Packing

A variation in the general DNA fiber packing arrangement found in the chromosome arms occurs at the centromere. It appears that the last stage of chromatin condensation, the coiling and compaction of the 250

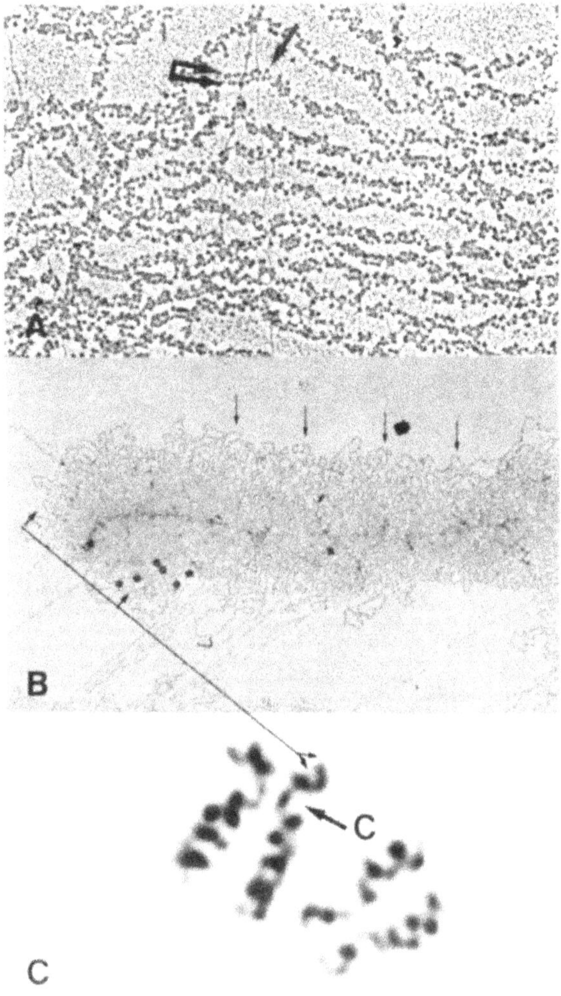

Figure 2. The higher order structure of metaphase chromosomes. **(A)** An electron micrograph of a series of slightly relaxed 30 nm chromatin fibers (bracketed arrows) from a human HeLa cell. Individual nucleosomes from the basic 10 nm fiber which form these structures are visible (arrow). **(B)** An electron micrograph of a partially condensed meiotic leptotene chromosome from the silkmoth *Bombyx mori* illustrating the tandem arrangement of a 30 nm fiber loop as a level of chromosome packaging. **(C)** A light micrograph of two partially relaxed human chromosomes illustrating the coiled chromatid fiber that is composed of 30 nm fiber loops. Note the straight course of the chromatid fiber in the region of the centromere (c) (Figure 2C reproduced from Rattner and Lin, 1985a).

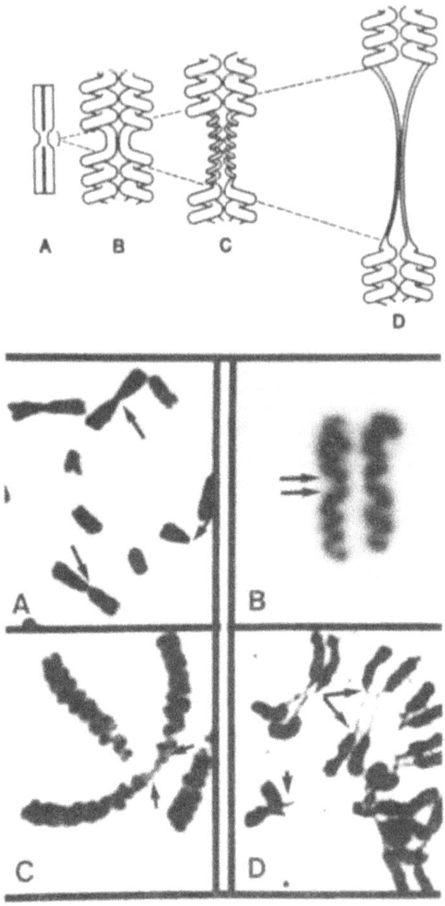

Figure 3. The higher order structure of the centromere. Diagram: A fully condensed metaphase chromosome is shown in A. The path of the chromatid fiber in the arms and centromere (bracketed area) are shown in B. The expanded centromere fiber seen in drug treated cells (see text) is shown in C and D (reproduced from Rattner, 1991).

Light microscope images of chromosomes displaying the morphology described in the preceding diagram. The micrographs are labeled to correspond to the diagram: **(A)** untreated fully condensed mouse chromosome. The centromeres are denoted by arrows, **(B)** partially relaxed human chromosome, **(C)** partially relaxed Indian muntjac chromosome. Note coils in the centromere region (double arrows), **(D)** 33258 Hoechst treated mouse chromosomes (arrows denote expanded centromere region) (reproduced from Rattner, 1991).

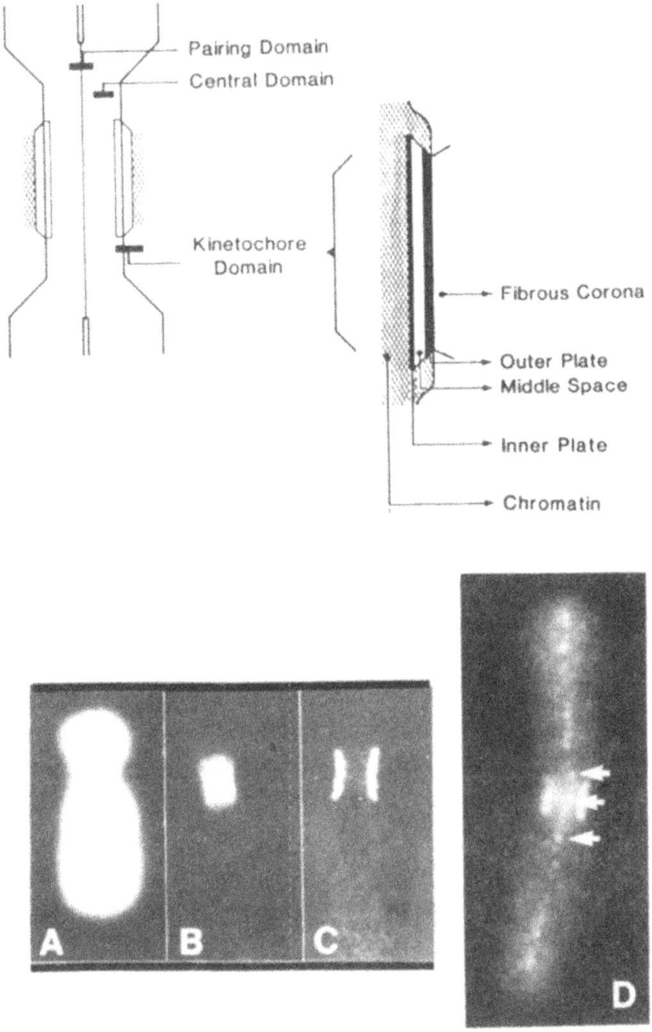

Figure 4. The domain organization of the kinetochore. Diagrammatic representation of the centromere. Figures A-D, identical Indian muntjac chromosomes: (**A**) whole chromosome reacted with a DNA specific stain reacted with antibodies to specific centromere domains, (**B**) whole centromere, (**C**) kinetochore domain, (**D**) arrows denote pairing domain [reproduced from Wong & Rattner (1992). The centromere. In: Advances in Molecular and Cell Biology, Bittar E.E., Ed. Vol. 4, pp. 1-35. JAI Press Inc., Greenwich].

nm fiber, does not occur at the centromere. Rather, the 250 nm fiber follows a straight course through the centromere (Figure 2C). This variation gives rise to the smaller diameter and constricted appearance of the chromatid in the centromere region.

The substructure of the 250 nm fiber which forms the centromere has been studied in some detail in the mouse *Mus musculus*. The centromeric heterochromatin of this species is largely composed of a highly repetitive AT-rich sequence known as the major satellite that is distributed throughout the centromere. When mouse cells are grown in the presence of either the drug 33258 Hoechst or 5-azacytidine that preferentially interact with AT-rich sequences, the normal condensation of the 250 nm centromere fiber is disrupted. The resulting extended centromere fiber is approximately five times the normal length of the centromere and with a reduced diameter of approximately 100 nm. Chromosomes prepared from cells recovering from drug treatment indicate that the extended centromere fiber can reform a centromere with normal morphology by coiling and condensation (Figure 3; for details see Rattner and Lin, 1985b).

There is a Relationship between Centromere Structure and Function

The multifunctional nature of the centromere has led to a search for a relationship between centromere structural organization and function. Protein mapping studies using antibody probes as well as ultrastructural studies have indicated that certain centromere functions map to spatial distinct regions of the centromere. To facilitate our discussion of the centromere, we can consider the centromere to be divided into three general domains: the pairing domain at the interface of the sister chromatids; the central domain, representing the interior of the centromere, and the kinetochore domain, representing the outer margin of the centromere (Figure 4).

The Pairing Domain. The pairing domain is found at the inner surface of the centromere and encompasses the centromere surface where sister chromatids interact. Two functions have been ascribed to this domain: first, to mediate the interaction between sister chromatids and, second, to ensure that sister chromatid separation occurs precisely at the onset of anaphase. Premature or incomplete separation will lead to the unequal partitioning of the genome and the formation of aneuploid cells. The nature of the interaction between sister pairing domains and the factors which affect this interaction are not clearly understood. There is some evidence to suggest that DNA in this region remains unreplicated until separation. Thus, DNA-DNA interactions may play a major role in sister chromatid interaction. In addition, several pairing domain specific proteins have been identified. These proteins have been given the designation CLIPs (Chromatid Linking Proteins) and INCENPs (IN CENtromere Proteins) and both localize to discreet patches at the surface of the paring domain. These patches may serve as the sites of linkage between the sister pairing domains implying that protein-protein interactions are also important in pairing domain function (for details see Cooke et al., 1987; Rattner et al., 1988). It should be noted that the CLIPs are also found along the pairing surface of the chromatid arms. Although early in division sister chromatids are associated throughout their length, the last point of attachment is at the centromere.

The Central Domain. The central domain is defined as the region between the inner and outer surface of the centromere and represents the bulk of the centromere. It has been suggested that the primary role of this domain is structural, since it is highly compact in nature and is composed largely of highly repetitive DNA sequences. The densely packed nature of this region may be required as a physical support for the formation and maintenance of the proper configuration of the pairing and kinetochore domains, as well as ensuring the proper spatial relationship between these two domains. In addition, since the centromere must absorb the mechanical force generated by the spindle on the chromosome, the features of the central domain may ensure the maintenance of chromosome integrity during chromosome movement. The DNA sequence and protein composition of the central domain centromere likely determines the unique higher order (constricted) structure of the centromere.

The Kinetochore Domain. Electron microscopy of metaphase chromosomes revealed that the outer surface of the centromere possessed unique surface specialization, the kinetochore (Figure 5A). Careful examination of conventional thin sections revealed that the kinetochore contains three layers: an inner electron dense layer known as the inner plate; a central electron translucent region known as the middle layer, and an outer electron dense layer, the outer plate (Figure 5B). When cells are treated with drugs, such as colcemid, a fibrous layer appears at the outer surface of the outer plate layer and has been termed the fibrous corona.

The inner and outer plates are roughly 30 nm in depth. Thin section and whole mount studies of intact and isolated kinetochores have revealed that the outer kinetochore plate is composed of a series of fibers ~30 nm in diameter. These fibers have a particulate substructure and exhibit properties that differ from that found for conventional chromatin. There is some evidence that this layer is rich in phosphorus and may be the site of phosphoproteins that have been detected at the kinetochore domain by antiphosphoprotein antibodies. Studies using antiDNA antibodies and DNA specific stains have detected DNA only in the inner plate and digestion of metaphase chromosomes, with nuclease dissolves the body of the chromosome while leaving the kinetochore intact (Figure 6). For review see Rieder, 1982; Brinkley, 1990.

The mammalian kinetochore domain functions as the point of integration of the chromosome with spindle microtubules. The kinetochore has the ability to capture spindle microtubules that grow from the spindle poles, as well as modulating the length of these microtubules. Thus, the kinetochore domain plays a fundamental role in chromosome movement (for review see Gorbsky, 1992). Microtubules interact primarily with the kinetochore outer plate, although there are numerous reports of microtubules inserting into the inner plate and the chromatin surrounding the margins of the kinetochore (Figure 5B,C). To date, mammalian DNA sequences which map to the kinetochore or function in specifying kinetochore placement remain unknown. However, a great deal of information has been obtained regarding DNA sequences that function like the mammalian kinetochore region in the yeasts.

Studies of the Yeast Centromere Revealed the First Details of the Molecular Organization of the Centromere

The primitive centromere found in the yeast *Saccharomyces cerevisiae* appears to differ in complexity from those found in mammals. For example, there is no evidence of the complex domain organization described above. Therefore, for *S. cerevisiae* the terms centromere and kinetochore can be considered synonymous.

The budding yeast *S. cerevisiae* was the first organism for which considerable detail about the molecular organization of the centromere was obtained. Using known genetic markers, chromosomal walking in an *S. cerevisiae* genomic library was preformed to isolate a functional centromeric sequence (CEN). When this DNA region was placed into an artificial minichromosome and introduced back into yeast cells, it was

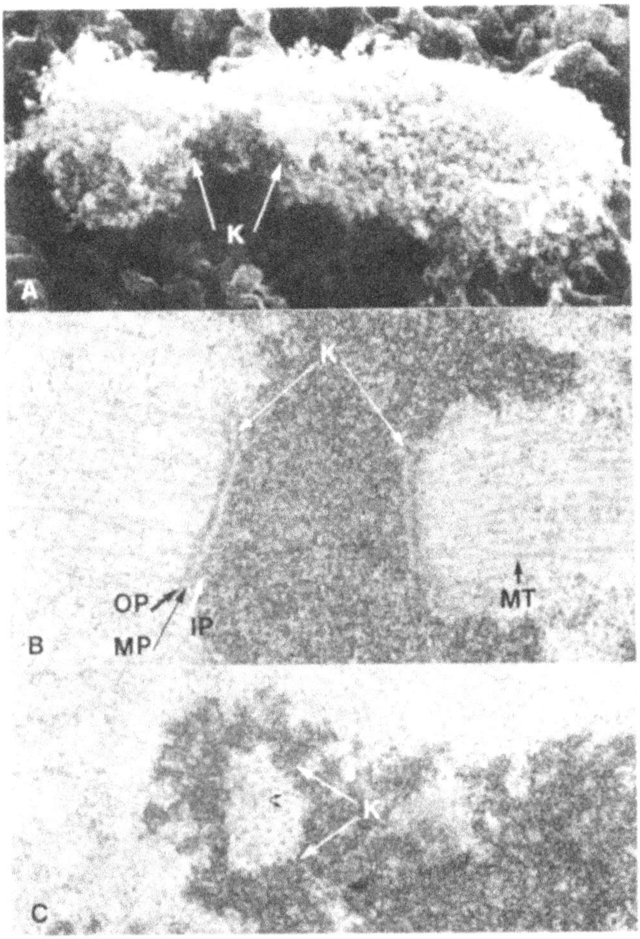

Figure 5. The kinetochore domain. (**A**) Scanning electron micrograph of an Indian muntjac chromosome showing the kinetochore (K) at the outer surface of the centromere. Note the fibrous substructure of the kinetochore outer plate. (**B**) Electron micrograph of a thin section through the kinetochore region of an Indian muntjac chromosome. Sister kinetochores (K) are seen on either side of the centromere. Each kinetochore is composed of an inner plate (IP), a middle plate (MP) and an outer plate (OP). The outer plate is, in turn, associated with microtubules (MT) emanating from the spindle poles. (**C**) An electron micrograph illustrating a kinetochore (K) region as seen in cross-section. Note the abundance of microtubules (small arrows) inserting into the kinetochore [Figure 5A reproduced from Rattner (1987). Chromosoma 95, 175-181].

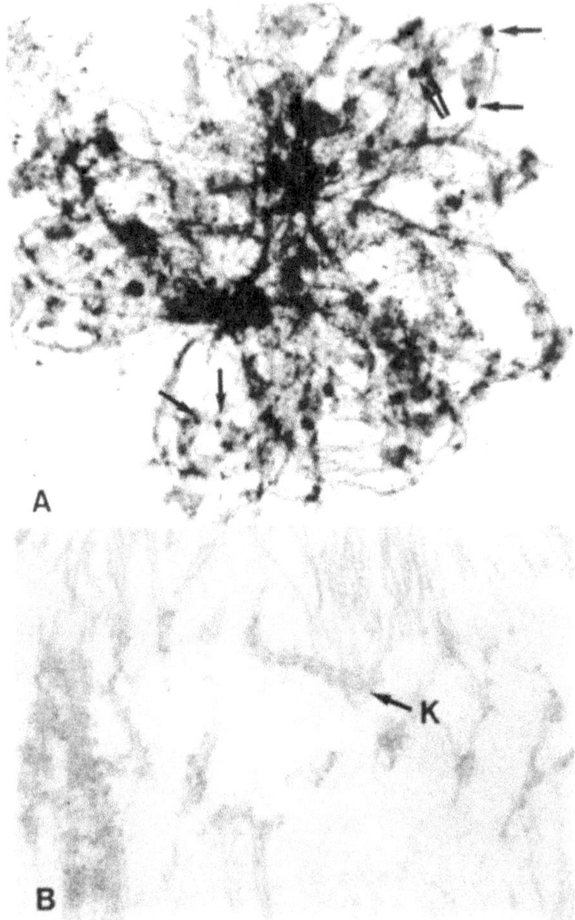

Figure 6. The kinetochore is more resistant to nuclease digestion than the remainder of the chromosome. (**A**) Electron micrograph of a whole mitotic spindle isolated after solubilization of DNA with the enzyme DNase I. Bundles of microtubules can be seen inserting into the kinetochores (arrows) which are the only remaining components of the chromosome. (**B**) Thin section of an Indian muntjac spindle after solubilization of the chromosomal DNA with micrococcal nuclease. Spindle microtubules can be seen inserting into the kinetochore which displays a typical tripartite structure (K). Note the absence of associated chromosomal DNA. [(A) reproduced from Rattner et al. (1978). Chromosoma 66, 259-268; (B) reproduced from Cooke et al., (1993). J. Cell Biol. (in press)].

shown to confer proper segregation and maintain mitotic stability. So far the CEN sequence in 13 of the 16 chromosomes of the *S. cerevisiae* karyotype has been identified. Each of these sequences has a common organization. Deletion analysis and DNA sequencing have revealed that there is a 125 bp functional region consisting of three distinctive domains. Element I consists of a consensus 8 bp motif PuTCACPuTG (where Pu is purine). Element II is composed of a variable 78-86 bp AT-rich region. Element III has a 25 bp conserved sequence that is partially palindromic: TGTTT(T/A)TGNTTTCCGAAANNNAAAAA. This latter element is absolutely essential for CEN function whereas elements I and II are only required for optimal activity. The chromatin configuration in a 220-250 bp region spanning these three elements is not nucleosome like, even though its flanking region is assembled into a highly ordered array of nucleosomes. Gel retardation assays and footprinting analysis have revealed that both elements I and III are protein binding sites. Three proteins (termed CP1 or CBF1) have been purified and are specific for element I. The CBF1 protein has been found to be crucial to the chromosome segregation process and also regulation of methionine metabolism (for review see Clark, 1990).

The Molecular Organization of the Centromere Differs between Species

The higher order organization of the chromosome and mitotic apparatus in budding yeast is quite different from that of the mammals. Budding yeast chromosomes do not condense and are not individually visible during cell division. The nuclear envelope does not break down and the spindle pole bodies are inserted within the nuclear envelope so that microtubules radiate within the nuclear space. Reconstruction of serial cross-sections of mitotic spindles suggests that one microtubule is attached to each daughter chromosome. Due to these differences, the focus of recent studies of centromere organization and function in the yeasts has turned to the complex fission yeast *Schizosaccharomyces pombe* which displays characteristics of higher eukaryotes. This species has a total of three large chromosomes (3.5, 4.7 and 5.5 Mb) which condense during mitosis. The centromeres in *S. pombe* are 300-600 fold larger than those found in *S. cerevisiae* and contain several moderately repetitive DNA sequences.

With genetic and molecular approaches similar to those used for *S. cerevisiae*, it has been possible to map the functional centromeric DNA sequences in all three chromosomes of *S. pombe*. These centromeric regions have been delineated on a 65, 100, and 150 kb restriction fragment for chromosome 1, 2, and 3, respectively. In general, each centromere consists of a central core of unique sequence DNA 4-7 kb in length flanked by a complex array of moderately repetitive, transcriptionally inactive sequences that behave as heterochromatin. It has been suggested that the central core sequence may serve as the attachment site for microtubules.

Do the structural and functional relationships which have been found in the centromeres of the yeasts also exist in higher eukaryotes, including the mammals? Unfortunately, at the present time we do not have enough molecular information about the mammalian centromere to answer this question. One possibility raised by recent studies is that the mammalian kinetochore may consist of multiple units, each of which are homologous in organization to the yeast microtubule insertion site (Zinkowski *et al.*, 1991).

The Sequence Composition of the Mouse Centromere Provides Insight into the Molecular Organization of Higher Eukaryotic Centromeres

Perhaps one of the most completely characterized satellite sequences is the major satellite of the mouse. It is estimated, based on reassociation kinetics, that there are at least 1 million copies representing 5-10% of the total DNA in the mouse, *M. musculus*. When the distribution of this sequence was first studied it became apparent that this sequence was a major component of all the centromeres of the karyotype except the Y chromosome and was distributed throughout the central domain as a series of a 234 bp repeating unit (Figure 7). The major satellite contains a curvature near the 3' end of the 234 bp monomer and the relaxation of this curvature with the drug dystamycin A interferes with the condensation of the centromere. Thus, the conformation of the DNA in the centromere may affect the condensation properties of satellite sequences and in turn may play a role in centromere structure. Interestingly, the general centromere distribution of the major satellite seen in *M. musculus* is not conserved throughout the mouse species, and in the older *Mus* species this sequence class is largely confined to the chromosomal arms. It is not clear, however, whether the major satellite exhibits the same properties and associates with the same proteins when it resides in different locations within the chromosome.

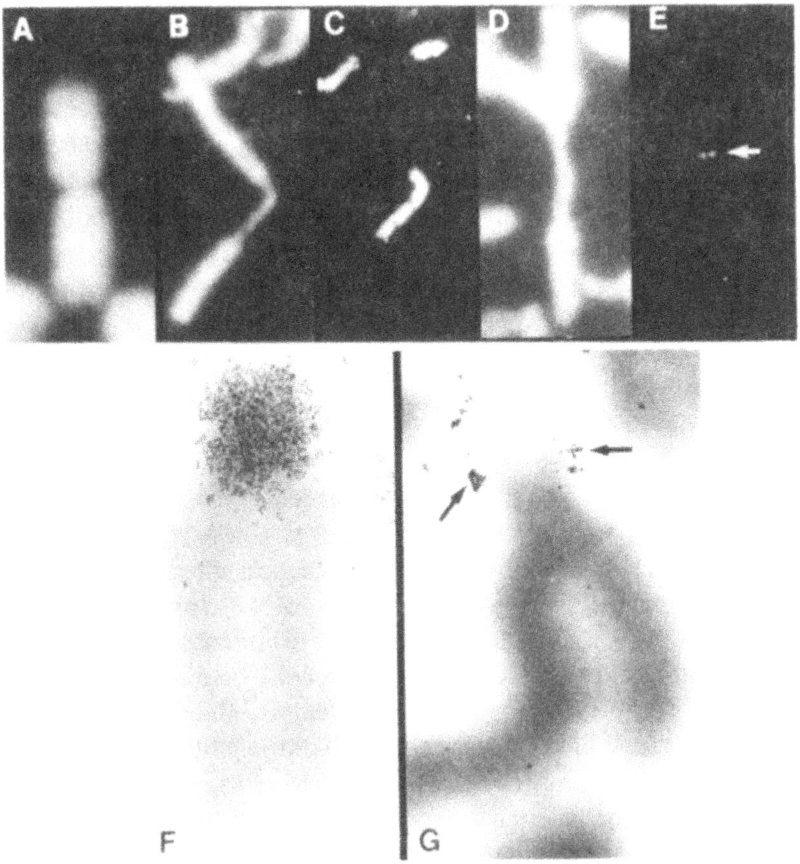

Figure 7. Distribution of DNA satellites in the mouse. (**A**) An untreated mouse chromosome stained for DNA. (**B-E**) *Mus musculus* chromosomes grown in the presence of 33258 Hoechst stained for DNA (**B, D**) and hybridized with either the major satellite (**C**) or the minor satellite (**E**). The general distribution of the major satellite throughout the centromere and the localization of the minor satellites to the kinetochore domain is also seen in electron micrographs of a whole mount *M. musculus* chromosome using an immunogold detection method. [Figures F and G courtesy of B. Hamkalo; Figures A-E reproduced from Wong and Rattner (1992). The centromere. In Advances in Molecular and Cell Biology (Bittar, E.E., Ed.) Vol. 4, pp. 1-35. JAI Press Inc., Greenwich.]

A second satellite sequence has also been found in all the centromeres of *M. musculus* except the Y centromere (Figure 7). This class, known as the minor satellite, has an abundance that varies within genomes and in *M. musculus* is found at a level of 50,000 copies per haploid genome. The basic monomer is 120 bp and sequence comparison suggests that the minor satellite has likely evolved from the major satellite through relatively recent evolution. Localization of the minor satellite within the centromere by *in situ* hybridization revealed that this satellite class co-localizes with the kinetochore domain in all the chromosomes of the karyotype, and immunoelectron microscopy has placed this sequence very close to the kinetochore plates (for review see Rattner, 1991).

Several Types of Human Centromere Sequences have been Identified

Three families of tandemly repetitive sequences have been identified in human chromosomes: classical satellites (I, II, III, and IV), alphoid sequences, and beta satellites. The four classical satellites have been mapped primarily to the heterochromatic long arm regions of chromosome 1, 9, 16, and Y. Each classical satellite contains a collection of different simple repeated sequences that are known as satellites 1, 2, and 3. For satellite 2 and 3, the repeating unit is essentially based on a degenerative form of the sequence (ATTCC)n. Satellite 1 is made up of two related AT-rich sequences, 17 bp and 25 bp arranged in tandem. The other satellite families, the alphoid sequences and the beta satellite, belong to the complex satellite class with a basic monomer repeating unit of approximately 171 bp and 68 bp, respectively.

The alphoid sequences account for 5% of the human genome and are found in every centromere of the human karyotype. Localization by *in situ* hybridization indicates that they are located primarily in the central domain. Blocks of these repeats have been sequenced and comparative studies have indicated that they are derived from alphoid sequences found in the African green monkey genome.

The intermonomer alphoid DNA sequence divergence within the human karyotype ranges from 20-40% and it exhibits a different hierarchical order in a given set of chromosomes. Subsets of the alphoid sequences are characterized by a multimeric higher order repeat unit. The copy number of the repeating unit tends to be the same between homologous chromosomes as opposed to nonhomologous chromosomes. The alphoid sequences, unlike the mouse satellites, have the ability to bind nonhistone chromosomal proteins including HMG 1 (for review see Willard, 1990).

Some Centromere DNA Sequences have been Identified that Specify the Binding of Centromere Specific Proteins

Several recent studies have indicated that a 17 bp segment, termed the CENP-B box, of the 171 bp human alphoid sequence binds to an 80 kD protein which has been identified as the centromere protein CENP-B (see below). A 19 bp region encompassing all the 17 bp CENP-B box has also been found in the recently evolved mouse minor satellites. Pairwise sequence comparisons of mouse satellites and alphoid sequences with other known centromere satellite sequences of unrelated eukaryotic species have revealed that a number of domains exist in the vicinity of this 19 bp motif with significant sequence identity. Taken together, these observations suggest that proteins associated with some centromeric sequences may have been retained throughout evolution and they may be related to conserved centromere structure and function (Masumoto et al., 1989; Sullivan and Glass, 1991).

The Mammalian Centromere Contains a Unique Subset of Chromosomal Proteins that are Present Throughout Cell Division

In concert with the identification of DNA sequences, there have also been major advances in the identification of proteins that reside within the centromere throughout the cell cycle (Table 1). The identification of these proteins has been facilitated by the discovery that many of these proteins can act as antigens in individuals with autoimmune disease. The most intensively studied centromere proteins are a group of antigens that have been designated CENP-A (17-19.5 kD), -B (80 kD), -C (140 kD), and CENP-D (50kD). CENP-A has recently been isolated, characterized, and shown to be a centromere-specific histone with some evolutionary relationship to histone H3. As indicated in the previous section, CENP-B has been found to bind to a specific DNA sequence located within the central domain. Recent evidence suggests that this protein has the ability to form dimers and thus may function in the folding and compaction of centromeric DNA (Yoda et al., 1992). CENP-C has been mapped to the kinetochore lower plate and also appears to be a DNA binding protein. CENP-D may be homologous

Table 1. Cell-Cycle Invariant Centromere

$M_r(kD)$	*Comments*	
14, 15, 15.5		Cox et al. (1983). Cell 35, 331-338; Aye
17m 18m 19.5	CENP-A; a histone H3 variant	Guldner et al. (1984). Clin. Exper. Imm mosoma 91, 313-321; Valdivia & Brinkl & Rattner (1987). Chromosoma 95, 403 815.
20, 23, 24		Cox et al. Cell 35, 331-338; Balczon &
50	CENP-D	Kingwell & Rattner (1987). Chromoson
70, 72		Nishikai et al. (1984). Ann. of Rheumat J. Immunol. 137, 2541-2547.
80	CENP-B	Earnshaw & Rothfield (1985). Chromos soma 94, 309-317; Kremer et al. (1988)
110		Earnshaw et al. (1984). J. Cell Biol. 98, 105, 855-862.
140	CENP-C	Earnshaw & Rothfield (1985) Chromos soma 96, 55-59.
11.7	HMG-I localizes to G/Q and C bands	Disney et al. (1989). J. Cell Biol. 109, 1
65, 135	Plant proteins	Molel-Bajer et al. (1990). Proc. Natl. A
170	Topoisomerase II (is not limited to the centromere)	Earnshaw & Heck (1985). J. Cell Biol.

Notes: CENP, centromere protein; HMG, high mobility group.

with the protein RCC1 which plays a role in chromatin condensation (for review see Rattner, 1991).

Microinjection studies have been performed using antibodies to CENP-A, -B, and -C in an effort to elucidate the function of these proteins. In tissue culture cells the antibodies are able to disrupt mitosis. Antibodies introduced 3 h prior to mitosis interfere with prometaphase chromosomal movement, while antibodies introduced during late G2 phase cause cells to arrest in mitosis. It has been suggested that in human cells the CENP antigens A-C are involved in two interphase events that are required for normal centromere function. This could include replication of the alphoid sequence and the structural maturation of the kinetochore at the beginning of prophase. Microinjection of CENP-A-C antibodies into mouse oocytes that are naturally arrested at second meiotic metaphase, eggs at first mitotic metaphase, or immature oocytes at first meiotic metaphase does not affect anaphase chromosome separation. However, prometaphase chromosome movements are disrupted if the antibody is injected into meiotic and mitotic eggs during interphase or prometaphase. Also, the subsequent anaphase is aberrant. Taken together, these two sets of experiments indicate that the CENP proteins have some functions in the organization of the centromere and that these roles may begin prior to the onset of cell division (Earnshaw and Bernat, 1991).

Some Proteins Transiently Associate with the Centromere

Chromosomal protein studies have also identified a group of centromere proteins that are associated with the centromere in a cell cycle specific manner (Table 2). The majority of these proteins appear to localize to the kinetochore domain during cell division and include several ubiquitous cellular proteins. In addition, this group also includes a set of proteins known as chromosomal passenger proteins. Many members of this latter group are believed to use the chromosome as a vehicle for transport to the mitotic apparatus where they dissociate to function within the mitotic spindle.

Recently, two new transient kinetochore domain proteins with a CENP designation (E and F) have been described. CENP-E is a 312 kD protein that is not detected at the centromere until prometaphase, where it maintains its localization through metaphase. At anaphase it becomes redistributed to the interzone region of the spindle and incorporated into the intercellular bridge by telophase. Microinjection studies suggest

Table 2. Cell-Cycle Regulated Centromere

$M_r(kD)$	*Comments*	
*Group A: Kinetochore Domain-Associated**		
16.8	Calmodulin	Dedman et al. (1980). Calmodulin: Its role i Structure and Function (Siege, Carafoli, Kre Elsevier-North Holland, Amsterdam.
34	p34^{cdc2} Kinase	Rattner et al. (1990). Cell Motil. & Cytoskel.
48	Enolase or Enolaselike	Johnstone et al. (1989). J. Cell Biol. 100, abs
50	Tubulin	Pepper & Brinkley (1977). Chromosoma 60,
79, 420	HeLa cytoplasmic dynein	Pfarr et al. (1990). Nature 345, 263-265
312	CENP-E	Yen et al. (1992). Nature 359, 536-539
400	CENP-F	Rattner et al. (1993). Cell Motil. Cytoskelet
440	Chicken cytoplasmic dynein	Steuer et al. (1990). Nature 345, 266-268
Group B: Central Domain, Pairing Domain-Associated or Unspecified Domain Assignment		
36	MSA-36	Rattner et al. (1992). Chromosoma 101, 625
40	Metaphase specific	Balczon et al. (1990). J. Cell Sci. 97, 705-71
59	Metaphase specific	Hadlaczky et al. (1989). Chromosoma 97, 2

Table 2. (Continued)

135, 140, 155	INCENPs (pairing domain)	Cooke et al. (1989). J. Cell Biol. 105, 2053-206
140, 155		Pankov et al. (1990). Chromosoma 99, 95-101
180, 210	Centrophilin	Tousson et al. (1990). J. Cell Biol. 112, 427-440
N.D.	CLIPs (pairing domain)	Rattner et al. (1988). Chromosoma 96, 360-367 32, 73-78
Group C: Centromere Plus Other Chromosomal Sites		
38	Stembody protein	Kingwell et al. (1987). Cell Motil. & Cytoskel.
62	Metaphase specific scaffold protein	Fields & Sharper (1988). J. Cell Biol. 107, 833-
70	Topoisomerase I enriched at centromere before anaphase	*Maul* et al. (1986). Proc. Natl. Acad. Sci. USA
62	Metaphase specific scaffold protein	Fields & Sharper (1988). J. Cell Biol. 107, 833-
70	Topoisomerase I enriched at centromere before anaphase	*Maul* et al. (1986). Proc. Natl. Acad. Sci. USA

Notes: CLIPs, chromatid-linking proteins; INCENP, inner centromere protein.
*Some also found at other sites within the mitotic spindle. N.D., not determined.

that this protein is important for the progression from metaphase to anaphase and molecular cloning has revealed that CENP-E is the largest member of the kinesin superfamily of microtubule-based motors.

CENP-F is a 372 kD protein that appears in the interphase nucleus at G2, hours before the appearance of CENP-F. However, like CENP-E, this protein localizes to the kinetochore domain and redistributes to the spindle midzone at anaphase. The cell cycle specific pattern of these two proteins suggests that some transiently associated centromere proteins may also function in other regions of the mitotic apparatus (for details see Rattner, 1991; Yen et al., 1992). The movement of these proteins within the spindle may help to determine and maintain the sequence of mitotic apparatus associated events.

The Centromere and Disease

The multiple functions of the centromere and its complex composition provides the basis for the centromere as a factor in several types of human diseases. For example, in some instances the organization or function of the centromere is perturbed so that the centromere can no longer function properly in its role in chromosome movement. This altered function may lead to the failure of the chromosome to attach to the spindle. Alternatively, the chromosome may be able to capture spindle microtubules but not modulate their length. Thus, the chromosome cannot move to the poles at anaphase. In both cases there is an unequal segregation of the genome leading to aneuploidy. Aneuploidy is defined as the addition or loss of one or more chromosomes from the karyotype. Approximately 0.5% of newborn infants have chromosomal abnormalities of which up to 50% may be due to aneuploidy. Aneuploidy is frequently seen in miscarriages.

A clinical example of improper centromere function is found in the human autosomal recessive developmental disorder known as *Robert's Syndrome*. The cytological abnormalities associated with this disorder include premature separation of the centromere and the failure of some or all of the chromosomes to move to the poles at anaphase. Thus, in this disorder the proper function of both the pairing and kinetochore domain appears to be affected. In general, the formation of an aneuploid cell can lead to abnormal cell function or cell death. Aneuploid cells are found in association with many types of cancer.

As previously mentioned, the identification of many protein components of the centromere have been facilitated by the identification of corresponding autoantibodies in the sera of patients with autoimmune disease. While the relationship between these circulating autoantibodies and autoimmune disease is unclear, their presence has been of use in diagnosis. For example, as many as 90% of patients with the CREST (**C**alcinosis, **R**aynaud's phenomenon, **E**sophageal dismotility, **S**clerodactyly, **T**elangiectasia) variant of systemic sclerosis show the presence of at least one of the centromere proteins, CENP-A, -B, and -C. When purified or recombinant centromere proteins are used in a sensitive assay (e.g., ELISA, immunoblotting), it has been shown that antibodies to CENP-B are found in virtually all systemic sclerosis sera that have anticentromere antibodies. The clinical value of detecting anti-centromere antibodies in patients with Raynaud's phenomenon has been illustrated by observations that the presence of these antibodies in the serum can accurately predict which patients eventually develop systemic sclerosis.

TELOMERES

Terminology

Telomere

A protein DNA complex that is the end of a chromosome.

Telomere Structure and Replication

Telomeres have a unique structure

The end of the metaphase chromosome is called the telomere and represents the end of the coiled chromatid fiber as well as the end of the single DNA strand which extends along the length of the chromatid (Figure 8). Chromosomes without telomeres are generally subject to breakage, fusion, and eventual loss. Studies of chromosome organization from a wide range of organisms indicate that the form of telomere DNA is conserved between many higher eukaryotes although the precise sequence may vary (Table 3). In humans the telomere has the structure AGGGTT and there are 650-2500 copies of this sequence at the ends of each chromatid yielding a total length of 4-15,000 bp. A variation in the length of the telomere has been noted between tissue types. The telomere repeat shows a specific organization with respect to the end of the chromosome. Thus, the end of each chromosome has one

Table 3. Telemoreric DNA Sequences

Telomeric Repeat	*Group*	*Organisms*
AGGGtt	Mammals	*Homo sapiens*
	Slime molds	*Physarum, Didymium*
	Filamentous fungi	*Neurospora*
	Kinetoplastid protozoa	*Trypanosoma, crithidia*
GGGGTT	Ciliated protozoa	*Tetrahymena, Glaucoma*
GGG[GT]TT		*Paramecium*
GGGGTTTT		*Oxytricha, Stylonychia, Euplotes*
AGGGTT[TC]	Coccidial protozoa	*Plasmodium*
AGGGTTT	Higher plants	*Arabidopsis*
TTTTAGGG	Algae	*Chlamydomonas*
(A)$G_{2\text{-}5}$TTAC	Fission yeasts	*Saccharomyces pompe*
$G_{1\text{-}3}$T	Budding yeasts	*Saccharomyces cerevisiae*
$G_{1\text{-}8}$A	Cellular slime molds	*Dictyostelium*

Notes: Sequence repeat unit shown 5' to 3' is the 3' end strand of the chromosome [reproduced from Blackburn (1991). TIBS 16, 378-381]. Square brackets have been used when either nucleotide may be in the motif. Round brackets indicate that the nucleotide may or may not be present.

G-rich strand which is oriented 5' to 3' towards the end of the chromosome and overhangs, by 12-16 bases, the complementary C-rich strand. This region acts as a primer for the enzyme telomerase. Nucleosomes are not found within the telomere.

Telomerase is Used in the Synthesis of Telomeric DNA

Telomerase, a specialized form of reverse transcriptase, is a ribonucleoprotein enzyme that was first identified in the ciliate *Tetrahymena* and has now been found in several other organisms including human HeLa cells. The RNA moiety contains a C-rich sequence that acts as a template for the synthesis of the G-rich telomeric DNA strand (Figure 9A). The major function of telomerase is to prevent the progressive shortening of the ends of the chromosome that would occur as a result of semiconservative replication (Figure 9B). It has been suggested that the presence of low telomerase activity, or its absence, may lead to the loss of telomeric DNA which in turn could eventually lead to chromosome instability and contribute to cell aging,

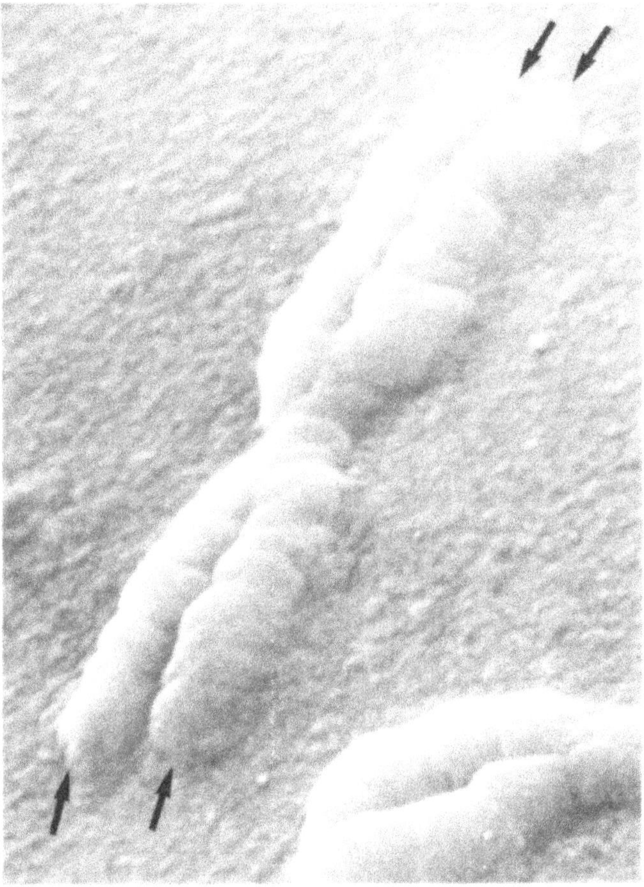

Figure 8. The telomere. A scanning electron micrograph of a human metaphase chromosome illustrating the position of the telomeres (arrows).

senescence, or even malignancy (for review see Blackburn, 1991; Greider, 1991).

In addition to dealing with the problem of maintaining the end of a chromosome, the structure of the telomere may also serve to prevent the end of the chromosome from fusing with those of other chromosomes. Studies on the organization of the chromosomes within interphase nuclei suggest that the telomere may associate with the nuclear envelope during a portion of the cell cycle. These functions raise the possibility that a unique subset of chromosomal proteins may localize to the

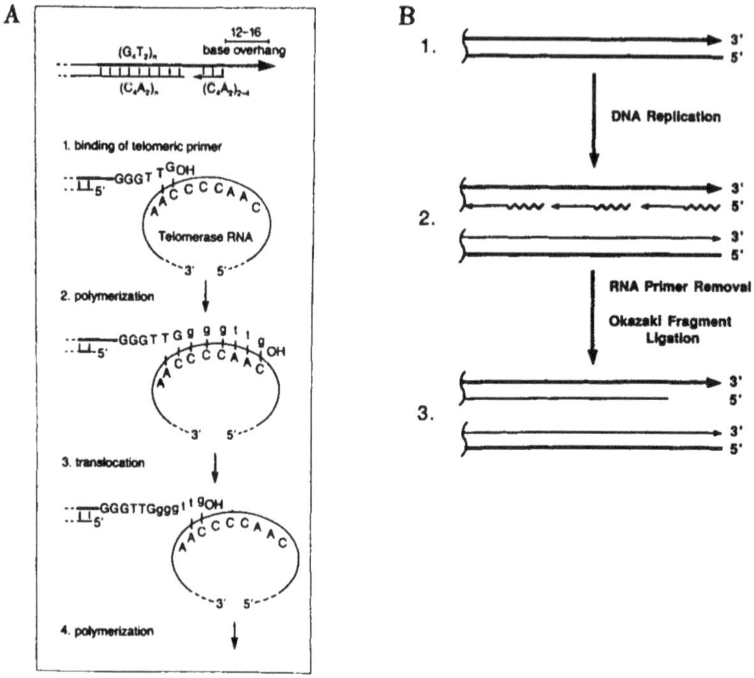

Figure 9. **(A)** Telomere structure and synthesis by telomerase. (a) Structure of the telomeric DNA of Tetrahymena. The G-rich strand (thickest horizontal line) and the complementary C-rich strand (heavy lower line) are in a duplex form except for the terminus (right), where the G-rich strand protrudes in a 12-16 base 3′ overhang. One or more single strand breaks occur in the distal part of the C-rich strand. (b) Mechanism of synthesis of the G-rich strand of telomeric DNA by telomerase. (1) The 3′ end of the telomeric DNA base-pairs with the complementary nucleotide in the CAACCCCAA template in the telomerase RNA. (2) Elongation of the telomere by polymerization of dGTP and TTP (indicated by lower case letters), copying the RNA template. (3) Translocation of the newly elongated 3′ end and repositioning of the template before (4) another round of polymerization (reproduced from Blackburn, 1991).
(**B**) The incomplete replication problem. (1) The molecular end of a DNA molecule is shown. (2) Leading strand replication proceeds to the end of the DNA molecule, while lagging strand replication utilizes RNA primers and Okazaki fragment synthesis. (3) Removal of the RNA primers and the Okazaki fragment ligation leaves the region at one end of the daughter molecule unreplicated. If there is no mechanism to fill the gap, the chromosome will get shorter with each round of replication [reproduced from Greider (1990). BioEssays 12, 363-369].

telomere. To date only two telomere proteins in the protozoan *Oxytricha* have been identified; however, their function remains ill-defined.

SUMMARY

The mammalian chromosome consists of several structural/functional regions the most prominent of which are the centromere and telomere. The centromere is a multifunctional chromosomal domain which plays a major role in chromosome segregation during cell division. This chromosomal region can be divided into three general domains based on structure and function: the pairing domain, the central domain, and the kinetochore domain. The molecular organization of the centromere is known in some detail for primitive organisms like the yeasts and in a more general way for higher eukaryotes. Abnormalities in centromere function lead to an unequal partitioning of the genome which can have a profound effect on cell viability and behavior.

The telomere, representing the end of the chromosome, consists of tandemly repeated simple DNA sequences. Telomeric DNA is copied from an RNA template that forms part of an enzyme known as telomerase. The unusual structure of the telomere and its associated proteins ensures the maintenance of the end of the chromosome, prevents chromosome-chromosome associations and may mediate chromosome-nuclear membrane interactions. Changes in telomere structure may affect chromosome stability, cell aging, senescence, and perhaps malignancy.

The identification of yeast centromere and telomere sequences along with sequences that serve as origins of DNA replication have led to the successful construction of a Yeast Artificial Chromosome (**YAC**). Since YACs can carry large amounts of DNA from other species, they have been the key to providing an avenue for isolating and mapping the human genome. Current research is focusing on the identification of human centromeres and autonomously replicating sequences (ARS) in an effort to create human artificial chromosomes which could be used to correct genetic defects.

REFERENCES

*Blackburn, E. H. (1991). Telomeres. TIBS 16, 378-381.

Boy de la Tour, E. & Laemmli, U.K. (1988). The metaphase scaffold is helically folded: Sister chromatids have predominantly opposite helical handedness. Cell 55, 937-944.

Brinkley, B. R. (1990). Views and reviews: Toward a structural and molecular definition of the kinetochore. Cell Motil. Cytoskel. 16, 104-109.

*Clark, L. (1990). Centrosomes of budding and fission yeasts. TIG 6, 150-154.

Cooke, C.A., Heck, M.M.S., & Earnshaw, W.C. (1987). The INCENP antigens: Movement from the inner centromere to the midbody during mitosis. J. Cell Biol. 105, 2053-2067.

Earnshaw, W.C. & Bernat, R. L. (1991) Chromosomal passengers: Toward an integrated view of mitosis. Chromosoma 100, 139-146.

Gorbsky, G. J. (1992). Chromosome motion in mitosis. BioEssays 14, 73-80.

Greider, C. W. (1991). Telomeres. Current Opin. Cell Biol. 3, 444-451.

Manuelidis, L. & Chen, T. L. (1990). A unified model of eukaryotic chromosomes. Cytometry 11, 8-25.

Masumoto, H., Masukata, H., Muro, Y., Nozaki, N. & Okazaki, T. (1989). A human centromere antigen (CENP-B) interacts wit a short specific sequence in alphoid DNA, a human centromere satellite. J. Cell Biol. 109, 1963-1973.

*Rattner, J. B. (1991). The structure of the mammalian centromere. BioEssays 13, 51-56.

Rattner, J. B., Kingwell, B., & Fritzler, M.J. (1988). Detection of distinct structural domains within the primary constriction using autoantibodies. Chromosoma 96, 360-367.

Rattner, J. B. & Lin, C. C. (1985a). Radial loops and helical coils coexist in metaphase chromosomes. Cell 42, 291-296.

Rattner, J. B. & Lin, C. C. (1985b). Centromere organization in chromosomes of the mouse. Chromosoma 92, 325-329.

Rattner, J.B., A. Rao, Fritzler, M.J., Valencia, D.W. & Yen, T.J. (1993) CENP-F is a ca400 KDa kinetochore protein that exhibits a cell-cycle dependent localization. Cell Motil. Cytoskeleton 26, 214-226.

*Rieder, C. L. (1982). The formation, structure and composition of the mammalian kinetochore and kinetochore fiber. Int. Rev. Cytol. 79, 1-58.

Sullivan, K. F. & Glass, C. A. (1991). CENP-B is a highly conserved mammalian centromere protein with homology to the helix-loop-helix family of proteins. Chromosoma 100, 360-370.

*Willard, H. F. (1990). Reviews: Centromeres of mammalian chromosomes. TIG 6, 410-416.

Yen, T. J., Li, G., Schaar, B. T., Szilak, I., & Cleveland, D. W. (1992). CENP-E is a putative kinetochore motor that accumulates just before mitosis. Nature 359, 536-539.

Yoka, K., Kitagawa, K., Masumoto, H., Muro, Y., & Okazaki, T. (1992). A human centromere protein, CENP-B, has a DNA binding domain containing four potential α helices at the NH_2 terminus, which is separable from dimerizing activity. J. Cell Biol. 119, 1413-1427.

Zinkowski, R. P., Meyne, J., & Brinkley, B. R. (1991). The centromere-kinetochore complex: A repeat subunit model. J. Cell Biol. 113, 1091-1110.

*Recommended Reading

Chapter 6

The Cytoskeleton

DAVID S. ETTENSON and AVRUM I. GOTLIEB

Principles of Medical Biology, Volume 2
Cellular Organelles, pages 121-145

ISBN:1-55938-803-X

INTRODUCTION

In this chapter, we will discuss the structure-function relationships of two important cytoskeletal systems in eukaryotic cells—actin microfilaments and microtubules. The third component of the cytoskeleton, intermediate filaments, will be discussed in Chapter 7. Many of the molecular concepts presented in this chapter are derived from *in vitro* biochemical studies of the various isolated proteins. Novel methods, however, are now allowing for cellular work to be done both in cell and organ culture systems and *in vivo*. Both the protein monomers actin and tubulin form long polymers by helical polymerization which are able to self-associate to form actin microfilaments and microtubules. The efficient methods of polymerization and depolymerization of these proteins make them well designed to provide mechanical support and integrate numerous cellular processes. The cytoplasmic cytoskeleton is involved in the regulation of cell shape, cell motility, mitosis and cytokinesis, intracellular transport, and axonemial growth and motility. An important biochemical function of these polymers is to provide energy for cell contraction and for cytoskeletal polymerization and depolymerization. Numerous associated proteins regulate polymerization-depolymerization, and binding of cytoskeletal proteins to self, other proteins, cell membranes and organelles. Thus the actin and tubulin monomers can be rapidly assembled into polymers in specific parts of the cell to carry out important cell functions. These cell functions can be further regulated by associated proteins which are able to alter the organization of the polymers. It is thus likely that cytoskeletal pathology does occur when the structure of the fibrous proteins are altered to produce abnormal cell functions which can lead to reversible and/or irreversible cell injury and cell death.

ACTIN—AN UBIQUITOUS MOLECULE

The actin monomer is a 42 kD globular protein composed of a single polypeptide 375 amino acids long. Actin is encoded by a small multigene family and its structure is very well conserved across eukaryotic phyla suggesting that the function of the molecule is important. Actin is a contractile protein which regulates the contraction of all types of muscle including skeletal, cardiac, and smooth muscle. It was a very exciting biological discovery that non-muscle cells which are not involved in the type of contraction produced by muscle were found to contain actin. This led to investigations on the structure and function of actin as it relates to nonmuscle activities and to the idea that non-muscle cells also contract, but for different purposes than do muscle cells. Actin is the most common protein in the cell cytoplasm and about 5% of the cells' total protein is actin in nonmuscle cells.

Vertebrate actins exhibit greater than 93% amino acid sequence homology. Even across eukaryotic phyla actin is highly conserved. Many of the functions carried out by actin are interchangeable between actins isolated from across many sources including yeast, drosophila, *Caenorhabdites elegans*, red blood cells, platelets, and numerous other tissues and cells. This suggests that the structure of the molecule has been maintained during evolution because it suits the very important functions that are carried out by actin polymers.

There are 3 isoforms of actin, α, β, and γ actin, each having a unique primary amino acid sequence. The β and γ-isoactins are the major forms in nonmuscle cells. There are also a series of muscle-specific isoactins, α-skeletal actin, α-cardiac actin, and both an α- and a γ-smooth muscle actin (Rubenstein, 1990). Have the isoactins evolved to accommodate specific functions? Since only a few residues vary among the isoforms and most of the replacements are chemically conservative, it is unlikely that fundamental functions are disrupted.

The type of actin isoform present in a particular cell type can change as conditions are altered. For example, in the adult artery wall, smooth muscle cells express predominantly α-smooth muscle (α-SM) actin isoform. However during development, in pathological states and in cell culture, the β-actin is predominant. After arterial injury, α-SM actin mRNA is decreased and the isoform decreases a few days later. Differences in function between different isoforms is not well understood. The role that these changes play in atherosclerotic plaque formation and restenosis following angioplasty are unknown.

There are also actinlike genes being identified. In some the core domain is homologous to actin while divergence occurs in other binding domains of the molecule. Some of these proteins are likely to be regulatory in function. An example of such a protein is centractin. Centractin is associated with centrosomes and may link the centrosome to actin or other proteins. Centractin displays over 50% homology with actin DNA sequences and more than 70% homology when conservative amino acid changes are considered. The regions corresponding to the core of actin are more conserved.

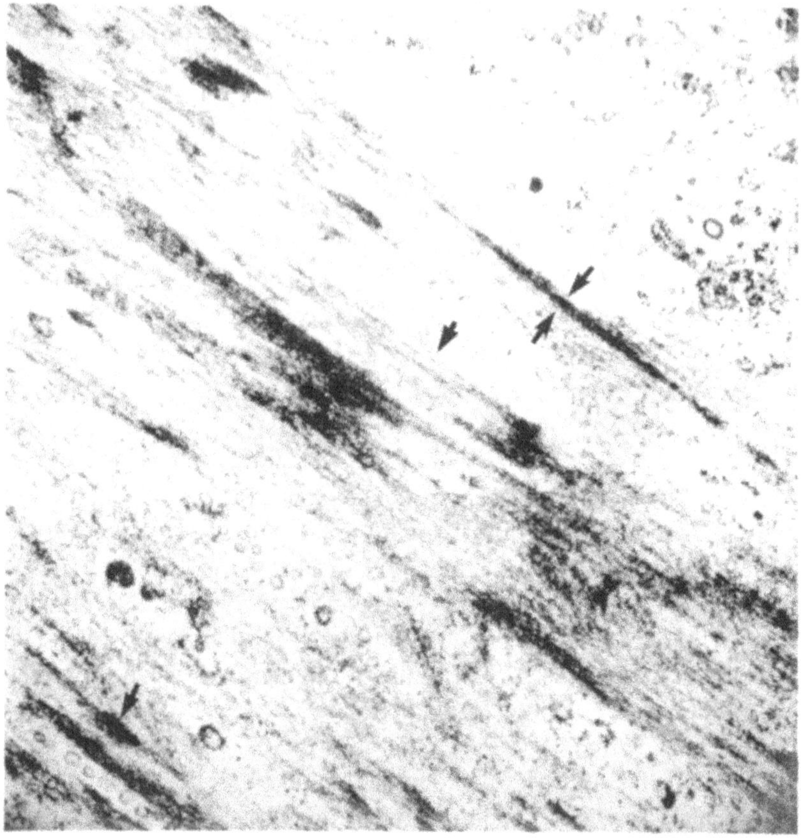

Figure 1. Electron micrograph showing actin microfilaments (arrows)

Formation of Actin Microfilaments

Actin monomers, referred to as globular actin or G-actin, self associate to initiate and then grow to form actin microfilaments (Figure 1), referred to as filamentous actin or F-actin (Bretscher, 1991). Two events are important in actin polymerization. The first is nucleation which occurs when 3 actin monomers come together in a specific geometric configuration. The second event is the addition of actin molecules to this initial complex. Assembly occurs at both ends of the polymer and requires ATP which is hydrolyzed to produce energy. Because actin microfilaments have structural polarity, the kinetics of polymerization are different at each end of the microfilament. Assembly at the plus end (barbed end) is more rapid than at the minus end (pointed end). This is due to changes in the conformation of each actin molecule as it attaches to the polymer. The conformational change has more of an effect at one end than the other. Usually monomers are added to the plus end and dissociate at the minus end. A steady state is achieved because polymerization is reversible. At this steady state, the assembly rate at the plus end is identical to the minus end and the polymer is kept at a constant length. Treadmilling is the term used to describe this process in which there is a flux of molecules but the polymer length stays the same. At steady state, the amount of polymerized actin will depend on the total actin present minus the free monomeric actin minus the actin bound to monomer-binding proteins. Cells have been shown to contain discrete foci of unpolymerized actin monomers (Fechheimer and Zigmond, 1993). These localized monomers of actin enable a cell, upon stimulation, to exhibit rapid polymerization.

While actin is distributed throughout the cytoplasm, most cells possess a dense network of actin microfilaments just beneath the plasma membrane. This network makes up the cell cortex, which gives mechanical strength to the cell surface and enables the cell to change shape and to move. Actin microfilaments also form bundles, some of which are stress fibers. These are composed of actin microfilaments in parallel alignment with nonuniform polarity.

Actin Binding Proteins

Actin binding proteins are proteins which bind to actin and in so doing alter the structure and function of the polymer (Stossel et. al, 1985). They are also important in determining the orientation and location of microfilaments. Some of the main classes of actin binding proteins are listed in Table 1. Since actin is highly dynamic it appears that the action of these proteins may be very rapid indeed.

Actin microfilaments participate in important cell processes when they are linked together to form bundles or networks. In skeletal muscle actin microfilaments are linked to myosin thick filaments to produce muscle contraction. Myosin is an ATPase and hydrolyzes actin ATP to produce energy for the sliding of myosin along actin. In the cortex of cells, actin microfilaments are often linked to the plasma membrane providing mechanical support to the cell membrane and support to

Table 1. Major Classes of Actin Binding Proteins

Name	*Molecular Weight (kD)*	*Function*
Myosin	260	• Involved in the sliding of actin filament for contraction • An ATPase
Tropomyosin	35	• Strengthens actin filaments • Promotes myosin binding
Gelsolin	90	• Ca^{++} dependent actin severing protein • Caps filament ends • Fragments actin network
Fimbrin	68	• Bundles actin filaments into tight bundles (10 nm separation)
α-actinin	200	• Bundles actin filaments in paral lel arrays (40 nm separation)
Filamin	27	• Cross-links actin filaments into gels • Two chains of filamin are linked head-to-head and their tails bind to actin
Profilin	15.2	• Sequesters actin monomers and prevents nucleation of actin monomers
Actin depolymerization factor	18.5	• Severs filaments • Promotes disassembly

cellular protrusions, either permanent ones such as villi or transient ones such as lamellipodia, filopodia and microspikes. Actin microfilaments in the cortex are likely to be involved in signal transduction pathways as well.

Defects in actin binding proteins can lead to pathological conditions. For example, a variant form of gelsolin can lead to the development of a familial amyloidosis (Finnish type) (McLaughlin, 1993). This variant of gelsolin has a single amino acid substitution which creates a site for a protease to fragment gelsolin. These fragments are found in the amyloid plaques. Another example is dystrophin, which is normally found in skeletal muscle and shows homology with α-actinin. It is localized at the plasmalemma and is required to anchor certain integral membrane proteins, such as dystrophin-associated glycoproteins, which may control calcium flux. Patients with the X-linked recessive diseases Duchenne and Becker muscular dystrophy lack dystrophin (Cox et al., 1993).

Microfilaments and Cell Adhesion

Stress fibers are examples of prominent bundles of actin microfilaments. They are usually found close to the sides of the cell that abut on the matrix and are indirectly attached to focal adhesion plaques (focal contacts). They have contractile function since they contain myosin filaments, tropomyosin, filamin and α-actinin. These bundles are attached to plaque structures that are associated with the inner aspect of plasma membrane domains. There are several nonmembrane proteins, associated with the focal adhesion including vinculin, α-actinin, radicin, and talin. The focal adhesion plaques are also associated with the transmembrane molecules called integrins. These transmembrane glycoproteins bind to matrix proteins such as fibronectin and collagen, on the outside of the cell. For example, the fibronectin receptor binds to fibronectin on the outside of the cell and on the inside it binds to talin, which binds in turn to vinculin which binds to other proteins which then bind to actin (Figure 2).

The stress fibers are thought to be important in mediating adhesion to the substratum, tension for contraction that is required for cell migration, and integrating intracellular processes with extracellular stimuli, especially those coming from nonsoluble matrix. This is one way in which the extracellular matrix can modulate cell function. This also allows intracellular elements to regulate matrix structure and function.

Stress fibers are also found in the belt like zonulae adhaerans (terminal web) where they are important in epithelial and endothelial cell-cell adhesion. The adhaerens junctional complex regulates permeability across sheets of renal tubular epithelial cells. The bile canaliculus of hepatocytes regulates canalicular luminal closure by contraction of a distinct circumferential pericanalicular microfilamentous belt analogous to the terminal web. The canalicular contraction is an energy dependent process. It is believed that cholestasis may be due to disruption of this process.

When endothelial cells form a confluent, contact inhibited monolayer *in vitro*, the periphery of the cell contains prominent circumferential microfilament bundles which we have termed the dense peripheral band (DPB) as well as shorter central microfilament bundles (Gotlieb et al., 1991). In low density culture, even in islands of endothelial cells where there is cell-to-cell contact, a DPB is not formed. It has been shown, using double labeling with immunofluorescence microscopy, that actin

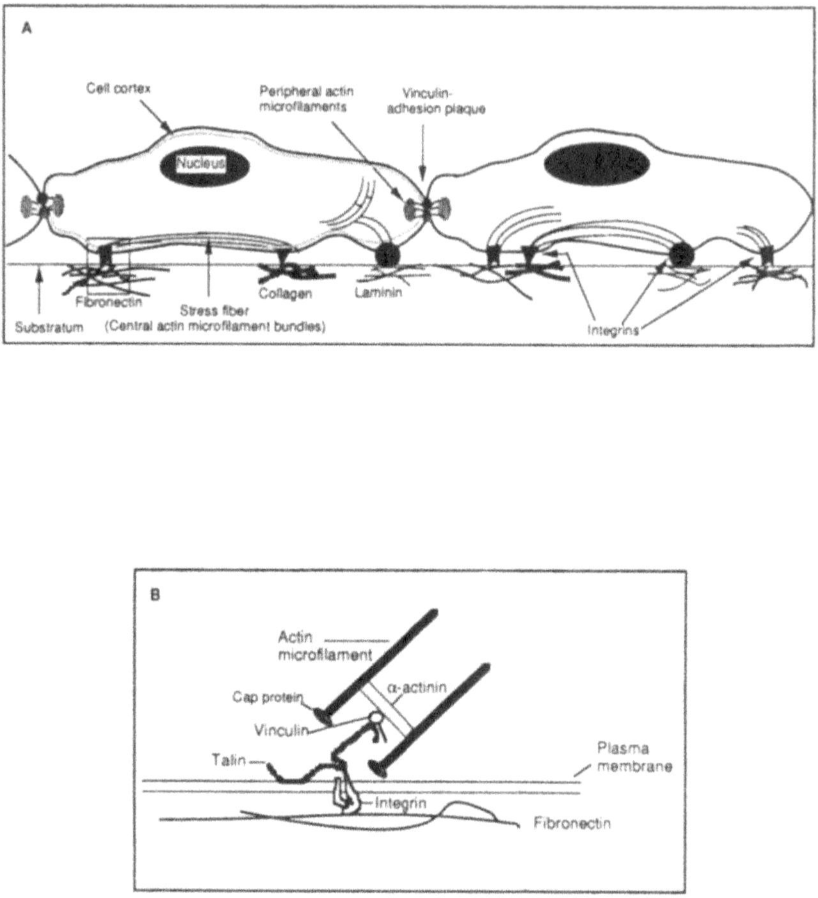

Figure 2. Schematic representation of (**A**) an endothelial monolayer showing cell-cell and cell-substratum adhesion. The cells adhere to each other via adhesion plaques that involve vinculin and the dense peripheral band of actin microfilament bundles. The cell binds to the matrix components through different integrins which interact with the microfilament network. (**B**) Details of some of the interactions that occur in the adhesion plaque.

colocalizes with myosin, with tropomyosin, with α-actinin, and with vinculin within this DPB. Although there are microtubules extending toward and into the DPB, there does not seem to be any preferential localization of microtubules within the band. Occasionally, microtubules run parallel to the band along its inner aspect. As noted above, *in vivo* actin microfilaments are also distributed as peripheral and central bundles. In the thoracic and abdominal aorta, the peripheral actin is less prominent than in confluent cultures. The central bundles are similar but are more prominent in the abdominal aorta. They are generally oriented in the long axis of the cell. However, deviations of up to a 30° angle are often seen. At areas where there is very low shear stress, the cells are cobblestoned and the morphology of the peripheral bands are similar to that seen *in vitro*. Transmission electron microscopic examination of the DPB has shown that there are microfilaments which emanate from the band and extend into junctions which have cytoplasmic plaques. Often the junctions of adjacent cells have microfilaments extending into their respective plaques and the microfilaments appear to be in alignment with each other. Since they have actin microfilaments extending into them and since vinculin is present at the periphery of the endothelial cells associated with the DPBs, it is likely that these plaques are similar to adhaerens junctions.

Thus in both epithelial and endothelial cells the stress fibers serve the very important function of keeping the cells together to form barriers and to line hollow tubes and spaces.

Microfilaments in Cell Contraction

Cell contraction occurs in skeletal, cardiac, and smooth muscle cells and in nonmuscle cells. Is there a single mechanism to explain cell contraction? This is suggested because associated proteins in nonmuscle cells are arranged on microfilament bundles in a pattern similar to that of a sarcomere in skeletal muscle. The orientation of actin microfilaments is aligned orderly in skeletal and cardiac muscle cells; however, in smooth muscle cells and more so in nonmuscle cells, the orientation is more random. In these cells, it appears that myosin regulates contraction. The molecule of myosin must be aligned correctly with actin to act as a ATPase and slide along the actin microfilament. One suggestion is that myosin molecules may assemble and bind to each other due to phosphorylation by myosin light-chain kinase which exposes the actin binding sites and allows the myosin tail to undergo a conformational change which allows the myosin to assemble into short bipolar filaments.

Microfilaments and Cytokinesis

In the late stages of mitosis a transient structure, the contractile ring, forms to eventually separate the daughter cells produced by cell division. This contractile ring contains actin and myosin and forms beneath the cell membrane at the middle of the cell and eventually disappears after it has contracted to pinch off the two daughter cells. The molecular mechanisms by which these contractile proteins are relocated to form this contractile ring at the cleavage site is not known.

Microfilaments in Response to Physical Forces

In vitro studies have demonstrated that flow-related shear stress alters several aspects of endothelial cell structure and function, including cytoskeletal organization.

The F-actin microfilament system, a component of the cytoskeleton, is important to the endothelial cell because of purported roles in the control of cell adhesion, cell migration, maintenance of cell shape, and cell permeability. *In situ*, F-actin is generally present as a continuous band around the periphery of cells and in microfilament bundles, or "stress fibers," in the central portion of cells. Some studies have reported that central microfilament bundles in endothelial cells are more numerous in regions of the arterial vasculature exposed to elevated shear stress. Kim et al. (1989) reported profound alterations in F-actin microfilament organization in endothelial cells at sites of elevated shear levels in the normal rabbit aorta. The peripheral microfilament band was disrupted, whereas the central stress fibers were markedly increased in thickness and length. The hypothesis that alterations in hemodynamic shear stress *in vivo* cause this reorganization of the F-actin microfilament system was supported by studies in which a 60% coarctation was used to alter shear stress in the mid-abdominal aorta of the rabbit. This procedure produces a region of moderately elevated shear downstream from the stenosis, in which peripheral F-actin was dispersed and large central stress fibers were formed. These findings indicate that *in vivo* actin microfilament distribution can be modulated by experimentally altering flow conditions. These findings are consistent with evidence of the influence of shear on F-actin distribution in cell culture systems.

The nature and speed of cytoskeletal changes were studied in the coarctation model by examining the restoration of a normal F-actin distribution after coarctations are removed and normal shear stresses are restored. We used abdominal aortic coarctation in rabbits to experimentally increase shear stress downstream from the coarctation by approximately twofold. *In situ* staining was employed to track subsequent F-actin redistribution. Within 12-15 hours, the number of stress fibers in the central regions of the cells decreased, and some separation of junctional actin in adjacent cells occurred. Long, central stress fibers of variable thickness were evident at 24 hours, but the band of actin normally seen at the periphery of the cells could no longer be distinguished. The redistribution of F-actin was completed over the next 24 hours by an increase in thickness of central stress fibers. Restoration of normal F-actin distribution after coarctations were removed proceeded more slowly. The long, thick stress fibers that were induced by high shear were replaced by thinner or shorter microfilament bundles 48 hours after the coarctation was removed. At 72 hours, central stress fibers were primarily long, thin structures. Peripheral F-actin was not fully restored until 1 week after removal of the coarctation, but there were still more and longer stress fibers at this time than were observed in control aortas. If current hypotheses linking central stress fibers to cell-substrate adhesion and peripheral actin to permeability regulation are correct, then our data indicate that induction of high hemodynamic shear stress may transiently compromise substrate adhesion. Subsequent alterations may enhance substrate adhesion, although intercellular permeability may be altered due to reductions in peripheral actin.

Microfilaments and Cell Migration

Endothelial cell migration occurs to repair areas of endothelial cell loss. This occurs on the surface of atherosclerotic plaques, in saphenous vein bypass grafts, and at angioplasty and atherectomy sites. The cells adjacent to the denuded area migrate when more than a few cells are lost. The migration has been studied *in vitro* and to a lesser extent *in vivo*; however, the latter studies show that *in vivo* findings are very similar to those described *in vitro*. *In vitro* studies showed the specific reorganization of microfilament bundles during aortic endothelial wound repair. As the cells at the leading edge elongated the DPB disappeared by 2 hours and central microfilaments were increased by 1 hour after wounding. Disruption of the microfilaments with a drug such as cytochalasin B results in a marked reduction in endothelial wound repair, indicating the importance of intact microfilaments for cell migration (Ettenson and Gotlieb, 1992).

Microfilaments in the Epithelial Cell Villus

The villus contains a rigid bundle of 20-30 parallel actin microfilaments which are oriented with the plus end towards the tip of the villus. Actin binding proteins including fimbrin bind the microfilaments to each other at regular intervals and there are lateral links formed between the actin microfilaments and the plasma membrane. The actin in the villus is not organized into stress fibers. There are no contractile functions noted. The actin microfilaments extend into the cytoplasm and terminate in the terminal web, which is a stress fiber. The terminal web is thought to stiffen the villi above it, keeping actin bundles projecting outward at a right angle to the apical cell surface.

Microfilaments and Secretion

Secretion requires the movement of secretory vesicles towards the plasma membrane which is under the control, in part, by the actin microfilament network in the cell. Most of the studies on the role of actin in secretion have been done in chromaffin cells (Trifaró et al., 1991). Under resting conditions actin controls chromaffin cell cytoplasmic viscosity through the formation of a mesh of actin microfilaments that are cross-linked and stabilized by fodrin and α-actinin. Both fodrin and α-actinin are actin binding proteins found in the secretory vesicle's membrane and serve as anchorage proteins. The cell plasma membrane and vesicle membrane also contain caldesmon, a calmodulin-dependent actin binding protein which, at low Ca^{++} concentrations, binds and cross-links actin microfilaments. When the cell is stimulated, Ca^{++} enters the cell and causes a dissociation of actin from fodrin and caldesmon, patching of fodrin along the plane of the plasma membrane, and activation of gelsolin with a consequent capping and shortening of the actin microfilaments. As a result, the cytoplasmic viscosity decreases, allowing the movement of vesicles towards the plasma membrane release sites where exocytosis occurs.

Microfilaments and Phagocytosis

Actin microfilaments play a major role in phagocytosis. It is known that actin microfilaments accumulate beneath the forming phagosomes and particle engulfment is blocked by cytochalasins, drugs that inhibit actin microfilament assembly. Whether the newly assembled actin microfilaments interact directly with the plasma membrane beneath the forming phagosomes is unknown. The formation of pseudopodia is due to actin polymerization but how this polymerization is translated into cellular movement is unclear. It is believed that newly formed actin microfilaments are quickly cross-linked by actin-binding proteins into an isotropic gel. This process is localized to the submembranous region resulting in a distending force that deforms the plasma membrane (Greenberg et al., 1991). The force causing pseudopod extension during phagocytosis is believed to be the result of elongation of the actin network.

Microtubules

Microtubules are cylindrical structures consisting of 13 protofilaments of tubulin each consisting of a heterodimer of α–and β-subunits, which are 50 kD polypeptides. These subunits are about 450 amino acids long and show about 40% homology. γ-Tubulin is an isoform associated with centrosomes and microtubule organizing centers. The microtubules provide a very important system of structures which can compartmentalize the cell cytoplasm and along which cytoplasmic organelles and membrane bound vesicles can be aligned and travel (Ludueña, 1993). The kinetics of assembly of microtubules is likely critical in organizing the distribution of microtubules and regulating cell shape, cell migration and cell division. Microtubules, although dynamic in nature, also render polarity to the cell.

The cytoplasmic microtubules emanate from the centrosome which is in the paranuclear area close to the Golgi apparatus (Figures 3 and 4). The centrosome contains amorphous material and the paired centrioles. It acts as a microtubular organizing center. This means that if all the microtubules are depolymerized in the cell by a microtubule- sensitive drug, then when the drug is washed out, the microtubules begin to appear and grow at the site of the microtubular organizing center. The fast growing end of the polymer is directed away from the centrosome toward the periphery of the cell. The rapidly growing end is referred to as the plus end and the slow growing end is the minus end. Hydrolysis of one molecule of bound GTP provides energy for polymerization and is important in producing a tighter binding of tubulin to the plus end than the minus end.

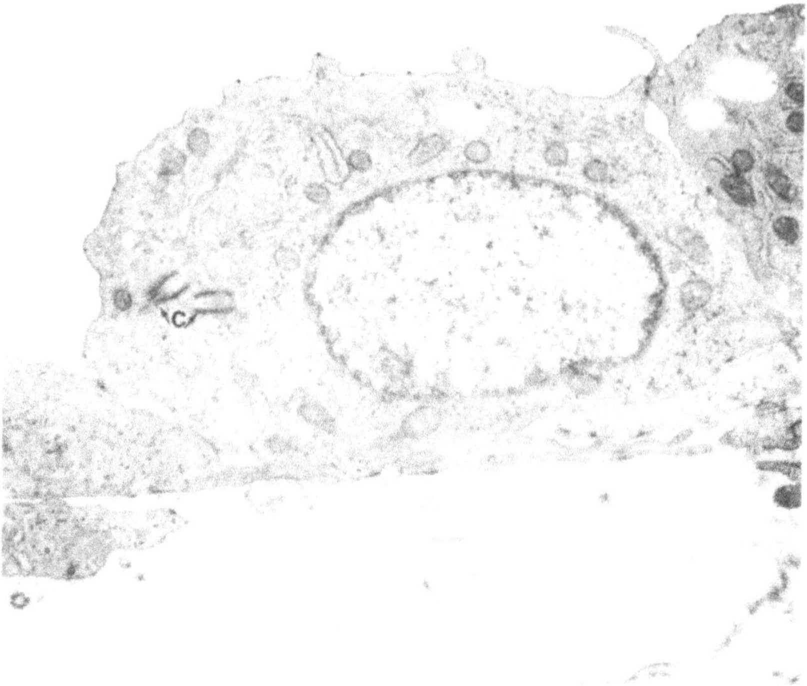

Figure 3. Electron micrograph of the centrosome, the paired centrioles (c), and the amorphous material around them in an endothelial cell.

It is thought that the dynamic instability of microtubules is an important mechanism by which the microtubules establish cell polarity and regulate the numerous functions attributed to microtubules. Dynamic instability refers to the fact that the ends undergo polymerization and then shrink only to extend again (Figure 5).

Thus, there is some randomness to the growth of the plus end of the polymer. If the plus end is stabilized, then depolymerization will stop. This stable microtubule provides polarization with respect to the direction and location of the peripheral microtubules and thus may regulate some specific function in that part of the cell periphery. More stable microtubules occur when enzymes present in the cytoplasm are able to acetylate or detyrosinate tubulin. Both the acetylated and detyrosinated tubulin are more stable, but there are cytoplasmic enzymes which are able to reverse the effect. In both cases it is the α-tubulin subunit that is affected.

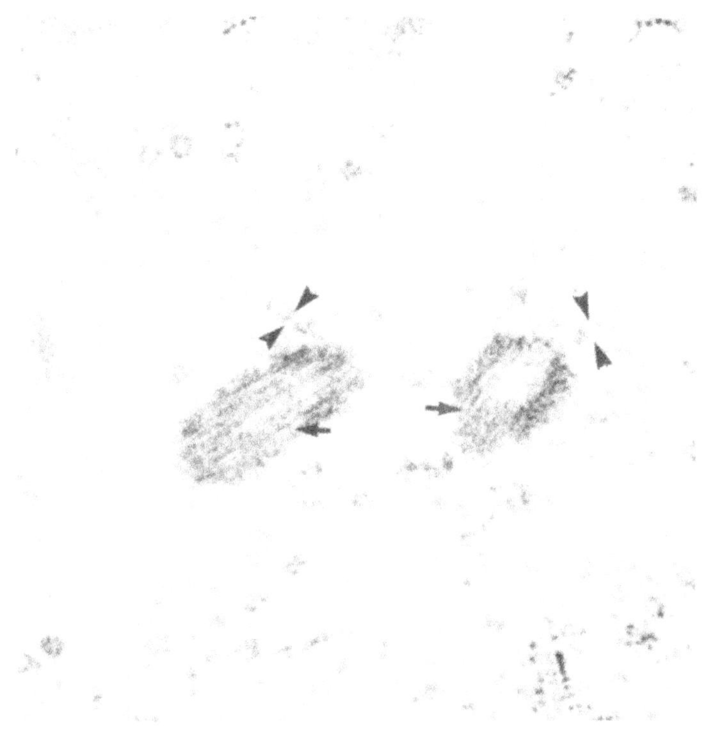

Figure 4. Electron micrograph of the centrioles (arrows) and microtubules (arrowheads) emanating out of them.

Associated Proteins

The microtubule associated proteins (MAPs) are thought to regulate the assembly and the interactions of microtubules. They bind to the highly negatively charged COOH-terminal region of microtubules and in so doing are able, with other binding domains, to stabilize the microtubules, link microtubules to cell organelles and vesicles, interact with other cytoskeletal elements, and act as motors to regulate intracellular motility. The high molecular weight MAP1 and MAP2, and the lower molecular weight tau proteins form two classes of associated proteins (Wiche et al., 1991). The high molecular weight MAPs are 200-350 in kD size. Tau comprises six isoforms of 48-67 kD in the human brain. MAP2 is abundant in brain; however, analogs have been found

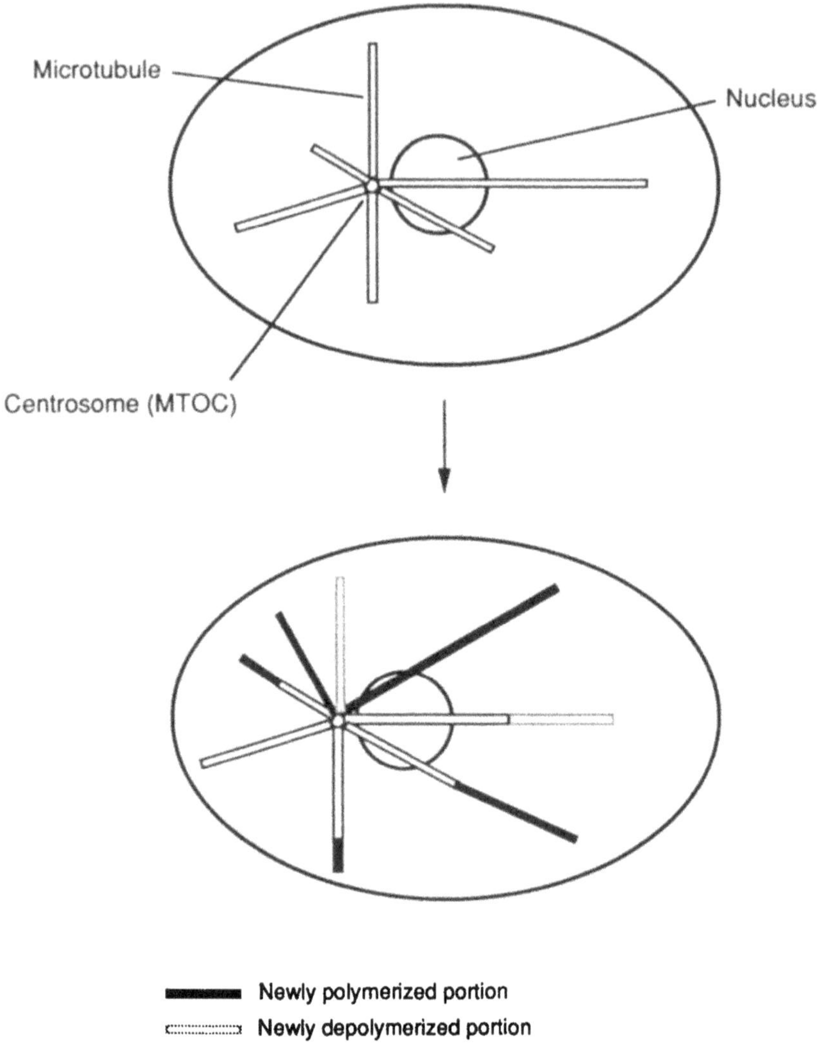

Figure 5. Schematic diagram illustrating the dynamic nature of microtubules. Tubulin subunits can rapidly assemble (black zones) and disassemble to lengthen or shorten the microtubule. Newly formed microtubules arise from the centrosome (circle).

elsewhere. There are several isoforms of MAP2 and the protein is present in neuronal cell bodies and dendrites and excluded from axons. The neuronal microtubule-associated proteins differ when comparing juvenile to adult forms. These difference often correlate over time with the establishment of stable neuronal pathways. This suggests that these proteins play an important role in growth and development and that specific microtubule-associated proteins are required to provide stability to the neuronal microtubules.

Tau proteins are a family of developmentally regulated phosphoproteins that are generated by alternative splicing of a single gene. Tau promotes assembly of microtubules and stabilizes them against depolymerization.

Microtubule Motors

Kinesin and dynein are two well-studied proteins that regulate intracellular movement along microtubules. Transport of membrane-bound vesicles and organelles in the axon of neurons is an important example of intracellular migration. In axons, the organization of microtubules is unidirectional with the plus end toward the synapse. Movement towards the plus end of the microtubules (anterograde direction) is due to the force generated by kinesin, while movement towards the minus end (retrograde direction) is due to dynein (Satir et al., 1990). These proteins coat the particle to be transported along the microtubules (Figure 6). Kinesin consists of two 124 kD heavy chains and two 64 kD light chains arranged so that two small globular heads protrude at one end of a coiled-coil stalk while the other end has small barbs. Each kinesin heavy chain possesses ATPase activity that resides in the head domains which bind to microtubules in an ATP-independent manner. The barb ends contain the ATP-independent region that binds to membrane-bound vesicles and organelles.

Dynein consists of two or three high molecular weight, heavy, polypeptide chains that possess ATPase activity, along with a number of intermediate and low molecular weight chains. The heavy chains reside in the globular head domain and bind to microtubules, while the other end binds to a membrane-bound vesicle or organelle and moves it towards the plus end of the microtubule. The kinesin or dynein coated particle is able to detach from one microtubule and reattach to another microtubule and thereby move to any destination in the cytoplasm.

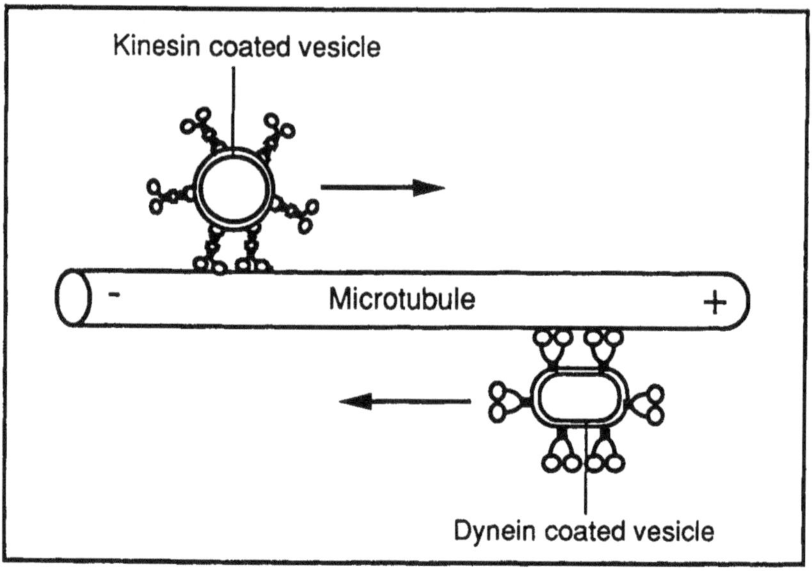

Figure 6. Schematic diagram showing the bidirectional movement of particles along a microtubule using kinesin to drive anterograde movement and dynein to drive retrograde movement.

Microtubules and Cell Migration

The centrosome plays a major role in the determination of cell polarity. Data from various biological processes including cytotoxic T-cell killing, mammary epithelial cells, and chemotaxis of neutrophils and macrophages bear out this relationship. In the specialized case of cell migration, several studies have shown that the position or redistribution of the centrosome determines the direction of cell movement in amoebae, 3T3 cells, fibroblasts and endothelial cells *in vitro,* and in aortic organ culture. It has been shown in endothelial cells that redistribution occurs before cell migration. Although it is accepted that the centrosome plays a role in the formation of microtubules, especially after depolymerization by colchicine-like compounds, the mechanism of interaction between centrosome redistribution and polarized cell activities, such as cell migration, is not well defined. Centrosomal redistribution in endothelial cells, however, can occur independent of cell migration. The presence of a wound edge, even if migration is inhibited, is enough to induce centrosome redistribution. It should also be noted that centrosomal redistribution requires intact microtubules and that there may be microtubule-microfilament interactions that enhance the redistribution. It has been recently suggested on the basis of morphological and immunocytochemical studies that microtubules in migrating fibroblasts select and stabilize focal long-lived contacts which function to nucleate the assembly of stress fibers, especially toward the leading lamellipodia.

In wound repair, the effector capacity of the centrosome is thought to be related to its close association with the Golgi apparatus. Using a transcribed viral protein as a probe for Golgi function, it has been shown that cell membrane insertion of this protein occurs preferentially in the area in front of the centrosome. Although biochemical function of the Golgi apparatus and the quantity of protein inserted into the membrane remained unchanged after disruption of the centrosome-Golgi apparatus association, polarized addition of the protein was impaired. Thus, the centrosome may play a central role of directing Golgi apparatus substances toward the cell periphery.

Centrosome redistribution does not occur during directional extrusion of lamellipodia to repair wounds of one to four cells in size (Gotlieb et al., 1991). Only when cell translocation was required in larger wounds did centrosome redistribution occur. Several implications arise from these observations. It is possible that centrosome-directed Golgi flow toward the membrane can occur in a manner independent of centrosome location by using preexisting microtubules which already extend toward the area of lamellipodia extrusion. An alternative explanation is that the directional extrusion of lamellipodia is an event independent of centrosome location. The cell membrane participating in cell spreading may come from the numerous blebs and folds found on the surface of cells. During cell translocation, however, active cell cycling becomes necessary because the membrane reservoir has been depleted by the preceding cell spreading. At present there is no direct evidence to support either concept.

Microtubules and Cilia

The basal bodies of cilia arise from centrioles. During ciliogenesis, the centrioles of the cell duplicate and migrate to a level just below the luminal border of the cell and become basal bodies. Like the centriole, the basal bodies are cylindrical structures with nine triplet microtubules in their walls. The two innermost microtubules of each triplet micro-

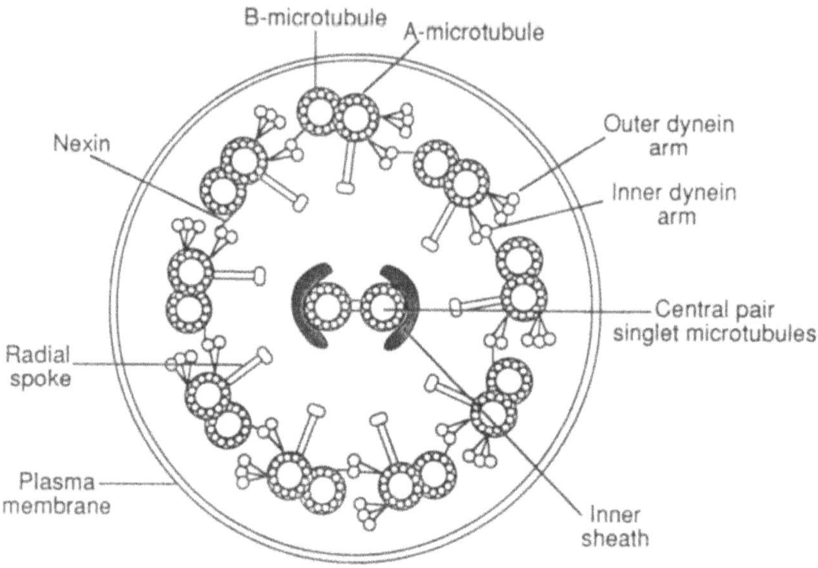

Figure 7. Diagrammatic representation of a cross-section of a typical cilium.

tubule grow out of the basal body and become peripheral microtubules of the cilium. The outermost microtubule of each triplet does not grow out of the basal body. Thus, the shaft of the cilium (axoneme) possesses nine peripheral doublet microtubules. In addition, two central singlet microtubules are present and are surrounded by a cell membrane (Figure 7). The nine outer doublet microtubules each consists of a complete, 13 protofilament microtubule (A-microtubule), and an incomplete, 10 protofilament microtubule (B-microtubule). These surround the central pair of single microtubules which are both composed of 13 protofilaments. Attached to the A-microtubules are pairs of dynein arms which project from the microtubule doublets and interact with adjacent doublets to produce ciliary bending. Projecting inward from each A-microtubule are radial spokes that are involved in regulating the form of the ciliary beat. In addition, adjacent doublets are held together by thin links of nexin which resist the sliding between adjacent doublet microtubules.

Cilia axoneme moves by a sliding microtubule mechanism. The dynein arms are responsible for this sliding. Dynein is a large protein complex containing two or three globular heads linked to a common root by thin flexible strands. Each globular head has an ATPase activity. These heads drive the sliding of microtubules in a cilium by a process that involves a regular cycle of conformational changes in each head. These changes are driven by ATP binding and hydrolysis which moves the dynein heads along a microtubule from its plus end towards its minus end. This movement generates force that drives the adjacent microtubule doublet toward the tip of the axoneme.

An understaning of the structure and function of normal cilia has provided information to identify defects in ciliary motility which can result in pathological conditions. For example, patients with Kartagener's syndrome (bronchiectasis, sinusitis, and situs inversus) have a defect in ciliary motility due to absent or irregular dynein arms. The lack of ciliary activity interferes with bacterial clearance, predisposing the sinus and bronchi to infection. This defect also affects cell motility during embryogenesis, resulting in situs inversus. Males with this syndrome are usually infertile due to ineffective mobility of the sperm tail.

Microtubules and Mitosis

The prominent microtubular cytoskeleton of an interphase cell becomes disassembled into its tubulin subunits and these subunits reassemble into two sets of polarized spindle tubules that become the mitotic apparatus.

During mitosis the centrosome microtubular organizing center (MTOC), divides and each set goes to opposite poles of the cell. The mitotic spindles originate and are attached to the centrosome. Spindles that extend from the poles and interact with the chromosome kinetochore are termed kinetochore microtubules. Polar microtubules extend from one pole to the other. Both types of microtubules are oriented with their plus ends away from the pole (Figure 8). The microtubules extend from the centrosome in all directions and only those that encounter the kinetochore are captured and stabilized. Although the kinetochore microtubules are firmly attached, they are still capable of disassembling and reassembling. This allows for the repositioning of the chromosomes during different stages of mitosis. The chromosomes are brought closer to the poles by the shortening of the kinetochore microtubules, while the polar microtubules lengthen to push the poles away from each other.

Microtubules and Axon Growth

Neuronal cells have a complex morphology in that most of their cytoplasm is extended over long distances in the form of axons. Axons

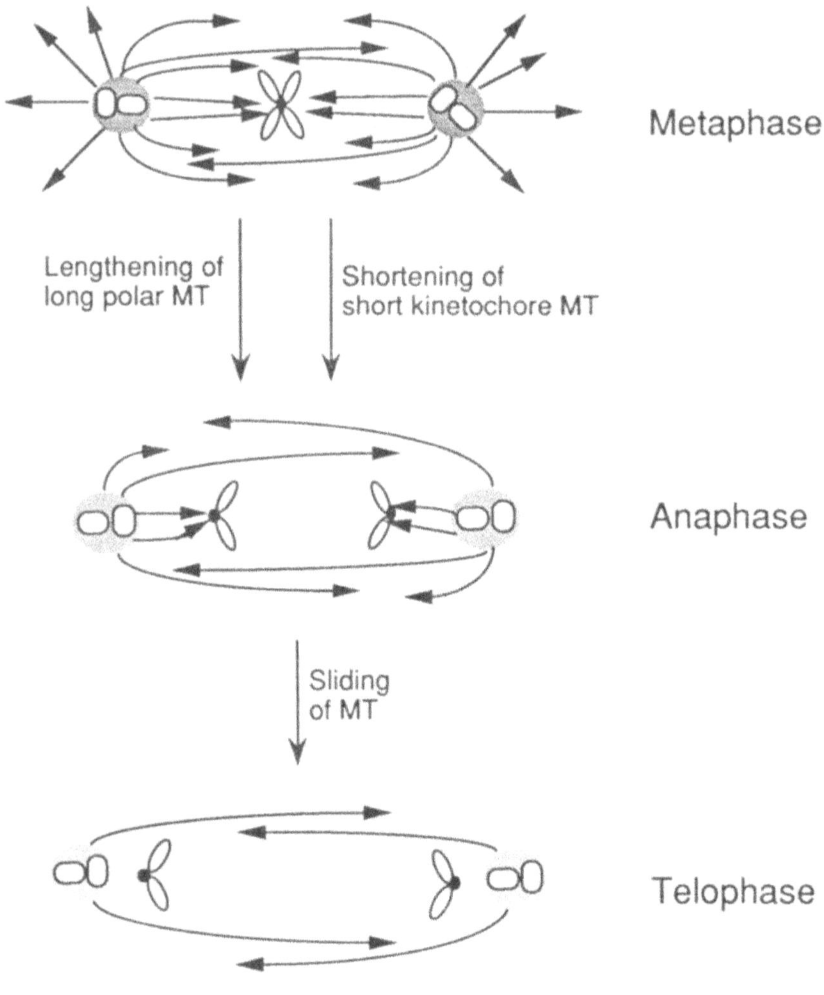

Figure 8. Schematic diagram showing the changes in microtubule length that accompany the movement of chromosomes during the different stages of mitosis. Only one set of chromosomes are represented. The polarity of the spindle microtubule is shown by the arrows (plus end).

are devoid of rough endoplasmic reticulum and Golgi apparatus and are thus dependent on the machinery of the cell body for producing the proteins required for their growth. As a result, axon growth requires an efficient process to transport proteins made in the cell body into and over great distances down the axon. In addition, axon growth is dependent upon mechanisms that generate and maintain the morphology of the cell. Microtubules provide both the structural support of the axon and direct the transport of organelles and proteins through the axoplasm.

Like microtubules in other cell types, axonal microtubules are uniformly oriented with their plus ends towards the periphery. However, unlike other cell types, axonal microtubules are not attached to a centrosome but are free in the cytoplasm (Joshi and Baas, 1993). It is believed that microtubules destined for the axon are initiated at the centrosome within the cell body, after which they are released from this site and are transported into the axon. The transport of these microtubules is unidirectional, thus establishing the polarity of the axon. The microtubules released from the centrosome are short and many elongate from their end as they move into and down the axon. Other microtubules shorten as they move down the axon, providing a source of tubulin subunits for the elongation of others. Along the length of the axon, individual microtubules have the ability to fragment in response to physiological stimuli, thereby rapidly increasing the local supply of microtubules as needed for events such as collateral branching. The microtubules that result from fragmentation inherit the centrosomally derived characteristics of their predecessor microtubules as well as their orientation. Thus, in this way, changes in the microtubule array leads to axonal growth.

Organelles and membrane-bound vesicles are transported along the microtubules by attaching to the mechanochemical proteins, kinesin and dynein, as discussed.

Alzheimer's Disease

The principal pathological features of Alzheimer's disease are senile plaques and neurofibrillary tangles. Paired helical filaments which are composed of the microtubule associated protein tau form the major fibrous component of the neurofibrillary tangle (Rubin et al., 1993). The tau proteins are abnormally phosphorylated in the diseased brain and this may be the reason that self-association leads to the formation of paired helical filaments. The abnormal phosphorylation may be due to enhanced kinase activity or reduced phosphatase activity which normally reverses phosphorylation. The concept is being actively studied by investigators. One suggestion is that the modification of tau noted above results in reducing the ability of it to nucleate and bundle microtubules. If this is so, then the defective organization of microtubules in the axons will result in abnormal axonal transport and abnormal synaptic transmission. This may lead to neuronal degeneration and eventual cell death.

SUMMARY

Actin microfilaments, microtubules and their associated proteins form complex filament networks that are very dynamic in nature. These cytoskeletal components play important roles in determining the shape and polarity of the cell and the nature of the cell's movement. These systems are also responsible for transporting material throughout the cytoplasm and they form an important pathway for signal transduction linking the cell cytoplasm to the external environment. Thus these filaments are important in maintaining the structure of the cell and regulating numerous cell functions.

REFERENCES

Bretscher, A. (1991). Microfilament structure and function in the cortical cytoskeleton. Ann. Rev. Cell Bio. 7, 337-374.

Cox, G.A., Cole N.M., Matsummura, K., Phelps, S.F., Hauschka, S.D., Campbell, K.P., Faulkner, J.A., & Chamberlain, J.S. (1993). Overexpression of dystrophin in transgenic mice eliminates dystrophic symptoms without toxicity. Nature 364, 725-729.

Ettenson, D.S. & Gotlieb, A.I. (1992). Centrosomes, microtubules, and microfilaments in the reendothelization and remodelling of double-sided *in vitro* wounds. Lab Invest. 66, 722-733.

Fechheimer, M. & Zigmond, S.H., (1993). Focusing on unpolymerized actin. J. Cell Biol. 123, 1-5.

Gotlieb, A.I., Langille, B.L., Wong, M.K.K. & Kim, D.W. (1991) Structure and function of the endothelial cytoskeleton. Lab. Invest. 65, 123-137.

Greenberg, S., El Khoury, J., Di Virgilio, F., Kapalan, E.M., & Silverstein, S.C. (1991). Ca^{2+}-independent F-actin assembly and disassembly during Fc receptor-mediated phagocytosis in mouse macrophages. J. Cell Biol. 113, 757-767.

Joshi, H.C. & Baas, P.W. (1993). A new perspective on microtubules and axon growth. J. Cell Biol. 121, 1191-1196.

Kim, D.W., Gotlieb, A.I. & Langille, B.L. (1989). *In vitro* modulation of endothelial F-actin microfilaments by experimental alterations in shear stress. Arterioslerosis 9, 439-445.

Ludueña, R.F. (1993) Are tubulin isotypes functionally significant? Mol. Biol. Cell 4, 445-457.

McLaughlin, P.J., Gooch, J.T., Mannherz, H.-G. & Weeds, A.G. (1993). Structure of gelsolin segment 1-actin complex and the mechanism of filament severing. Nature 364, 685-692.

Rubenstein, P.A. (1990). The functional importance of multiple actin isoforms. BioEssays 12, 309-315.

Rubin, G.C., lqbal, K., Grundke-Iqbal, I. & Johnson Jr., J.E. (1993). The organization of the microtubule associated protein tau in Alzheimer paired helical filaments. Brain Res. 602, 1-13.

Satir, P., Goltz, J.S. & Wolkoff, W. (1990). Microtubule-based cell motility: The role of microtubules in cell motility and differentiation. Cancer Invest. 8, 685-690.

Stossel, T.P., Chaponnier, C., Ezzel, R.M., Hartwig, J.H., Janmey, P.A., Kwaitkowski, D.J., Lind, S.E., Smith, D.B., Southwick, F.S., Yin, H.L. & Zaner, K.S. (1985). Nonmuscle actin-binding proteins. Ann. Rev. Cell Biol. 1, 353-402.

Trifaró, J.M., Vitale, M.L. & Castillo, R.D. (1992). Cytoskeleton and molecular mechanisms in neurotransmitter release by neurosecretory cells. Eur. J. Pharmacol. 225, 83-104.

Wiche, G., Oberkanins, C. & Himmler, A. (1991). Molecular structure and function of microtubule-associated proteins. Intl. Rev. Cytology 124, 217-273.

Chapter 7

Intermediate Filaments:
A Medical Overview

MICHAEL W. KLYMKOWSKY and ROBERT M. EVANS

Principles of Medical Biology, Volume 2
Cellular Organelles, pages 147-188

ISBN:1-55938-803-X

INTRODUCTION

Within the typical eukaryotic cell there is a fibrous "cytoskeleton" composed of microtubules, microfilaments, and intermediate filaments (IFs) (for general reviews see Schliwa, 1986; Bershadsky and Vasiliev, 1988; Bray, 1992). Microtubules act primarily as the tracks along which mechanoenzymes, e.g., cytoplasmic dyneins and kinesins, transport subcellular organelles. Microfilaments are also involved in cellular motility, both as tracks along which myosins move and through their controlled polymerization and network formation. In addition, the fundamental processes of mitosis, meiosis, and cytokinesis depend upon microfilaments and microtubules: both are essential for the survival of the eukaryotic cell. In contrast, IFs have not yet been directly associated with cellular or intracellular movements. In fact, IFs and IF subunit proteins (IFPs) do not appear to be present in many types of eukaryotic cells. For example, IFs have yet to be unambiguously identified in yeasts or other unicellular eukaryotes, although proteins analogous to IFPs may well exist (see McConnell and Yaffe, 1992, 1993). On the other hand, IFs have been found in a wide range of metazoans, most significantly from the human perspective, in the vertebrates (Bartnik and Weber, 1989). Moreover, in the vertebrates there has been an impressive increase in number of distinct IFPs over that seen in most invertebrates (see below). Analysis of protein sequence and gene structure indicates that the IFPs of vertebrates and invertebrates are related to one another as well as to the nuclear lamins, which form a fibrous layer (nuclear lamina) on the inner surface of the nuclear envelope (Aebi et al., 1986). Based on ultrastructural studies, nuclear lamina appears to be a ubiquitous and essential (Gerace and Burke, 1988) feature of eukaryotic cells; the nuclear lamins may well be the progenitor of the cytoplasmic IFPs (Weber et al., 1991). One plausible scheme for the evolution of IFPs is shown in Figure 1.

In vertebrates, it is clear that IFs are not strictly necessary for cell survival. There are cells that do not appear to express any IFPs, e.g., mature oligodendrocytes, the cells that myelinate axons in the vertebrate central nervous system (Choi and Kim, 1984; Ogawa et al., 1985). Plasma cells are also reported to lack IFs (Dellagi et al., 1983). Moreover, some tumors (Lilienbaum et al., 1986) and a number of cultured cell lines (Venetianer et al., 1983; Dellagi et al., 1984; Giese and Traub, 1986; Hedberg and Chen, 1986; Lilienbaum et al., 1986) have been found not to express IFPs. Finally, the experimental disruption of IF organization has no appreciable effect on the behavior of cultured cells (see Klymkowsky et al., 1983, 1989; Baribault and Oshima, 1991). On the other hand the conservation of IFP sequence and patterns of expression, particularly within the vertebrates, argues that IFPs are under some type of selection, presumably because a normal IF network is beneficial to the organism. Based on their phylogenetic distribution it seems that IFs are an evolutionary adaptation to a metazoan lifestyle in general, and the vertebrate lifestyle in particular. IFs are likely to play roles involving high levels of cellular organization. (See Klymkowsky, 1995).

Although the functional significance of IFs is not well understood, from a medical point of view IFPs are important in three ways. First, a number of diseases have been shown to affect IF organization and mutations in IFPs have been found to be the cause of some severe human diseases (see below). Given the number of distinct IFPs in humans, it is likely that the future will see the identification of additional human genetic defects as defects in IFs. Second, IFPs are expressed in a cell type-specific pattern that is generally preserved upon neoplastic transformation. IFP-specific antibodies are currently used to determine the origin of tumors that are ambiguous using less specific histochemical methods. And third, recent work has suggested that changes in IFP expression may correlate with metastatic potential in some tumor types (see Chu et al., 1993). Our goal in this chapter is to provide a broad survey of IFs and IFPs. We will discuss the IFPs of vertebrates, their expression, and organization in cells. We will also review what is known about the roles of IFPs in cells and tissues and their relation to human pathologies.

INTERMEDIATE FILAMENT PROTEINS

IFPs all share a characteristic structure (Figure 1) (Steinert & Roop, 1988). They are elongated proteins that have been divided into an N-terminal head domain, a central rod domain, and a C-terminal tail domain. This same structural organization is shared by the nuclear lamins. The rod domain consists largely of tandem repeats of a seven amino acid motif in which amino acids one and four tend to have hydrophobic side-chains. This heptad repeat is present in a number of fibrous proteins, including the tail of myosin (Cohen and Parry, 1986). There is a strong tendency for the heptad-rich rod domain to form an α-helix with a hydrophobic surface (Crick, 1953). These monomers dimerize to form coiled-coils. In the IFPs of invertebrates and in the nuclear lamins, the central rod domain is 352 amino acids long, while the rod domain of

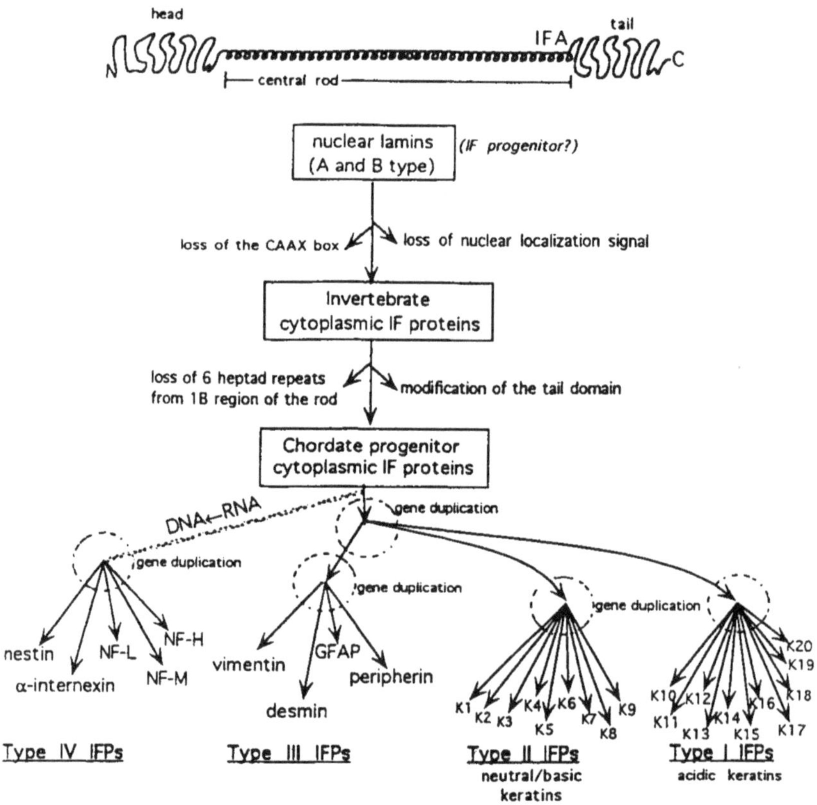
head
tail
IFA
N
C
central rod
nuclear lamins
(A and B type)
(IF progenitor?)
loss of the CAAX box
loss of nuclear localization signal
Invertebrate
cytoplasmic IF proteins
loss of 6 heptad repeats
from 1B region of the rod
modification of the tail domain
Chordate progenitor
cytoplasmic IF proteins
DNA←RNA
gene duplication
gene duplication
gene duplication
gene duplication
gene duplication
nestin
α-internexin
NF-L
NF-M
NF-H
vimentin
desmin
GFAP
peripherin
K1
K2
K3
K4
K5
K6
K7
K8
K9
K10
K11
K12
K13
K14
K15
K16
K17
K18
K19
K20
Type IV IFPs
Type III IFPs
Type II IFPs
neutral/basic
keratins
Type I IFPs
acidic keratins

Figure 1. A plausible scheme for the evolution of IF proteins. The conserved secondary structure of IF proteins is illustrated in the cartoon at the top of the figure. It consists of a central helical rod domain, flanked by nonhelical N-terminal head and C-terminal tail domains. The position of the highly conserved IFA epitope is marked. In the second part of the figure, the possible steps in the evolution of the cytoplasmic IFPs are shown (figure is derived from the work of Weber and colleagues). Assuming the progenitor of all IF proteins was a lamin, the first step was the loss of the nuclear transport signal and the "CAAX" box, involved in the isoprenylation of lamins (for review see Gerace and Burke, 1988). Next, came the loss of 6 heptad repeats from the central rod domain. The gene encoding this shortened cytoplasmic IF protein was duplicated to form the type III IF proteins. At some point, two side branches of this family of genes produced proteins that could no longer form filaments on their own, but were able to form filaments when present together. These two branches then underwent their own round of gene duplication to form the type I and type II keratins. Finally, in an event that appears to involve the reverse transcription of an mRNA, the neurofilament group of IF proteins (type IV) was founded. This founder gene underwent a period of gene duplication to produce the various neuronal IF proteins. During the course of this duplication, the tail domains of these proteins grew substantially. This scheme is not meant to be all inclusive. For example filensin, an IF-like protein present in the lens (Gounari et al., 1993) is not included. During its evolution it appears to have lost 29 amino acids from the central rod domain, together with its entire tail domain. How it fits remains to be determined (See Klymkowsky, 1995). More IF and IF-like proteins are sure to be discovered as well.

the vertebrate IFPs is shorter by six heptad repeats, with a length of 310 amino acids (Figure 1) (Weber et al., 1989, 1991). The rod domains of different IFPs are similar, particularly in a specific region at the C-terminal end of the rod domain. The monoclonal antibody antiIFA (Pruss et al., 1981), which recognizes almost all known IFPs, binds to this region of the protein. The head and tail domains of different IFPs, in contrast, are quite distinct. In particular, the tail domain varies in length from ≈11 amino acids to ≈600 amino acids depending upon the IFP. Mutational studies indicate that the rod and head domains of IFPs are critically important for IF assembly. The tail domain, on the other hand, appears to be involved primarily in the formation and organization of IF networks within the cell.

Vertebrate IFPs are a large family of proteins that have been classified into a variety of types (Figure 1). Based on analysis of genomic organization, it would appear that the type III IFPs, i.e., vimentin, desmin, glial fibrillary acidic protein (GFAP), and peripherin, most closely resemble the hypothetical progenitor of the vertebrate IFPs (Weber et al., 1991). Vimentin is expressed in a wide range of cell types and is often found to be expressed by differentiated cells when they are placed into culture. Desmin is expressed in smooth, cardiac, and skeletal muscle. GFAP is characteristic of astrocytic glia of the central nervous system, but has also been found expressed in glia of the peripheral nervous system. Peripherin is characteristically expressed in neuronal cells of the peripheral nervous system (see Fliegner and Liem, 1991 for review).

The largest group of IFPs are the keratins (see Steinert and Roop, 1988; Albers and Fuchs, 1992 for reviews). In humans, over 20 different keratins have been identified. Keratins are typically expressed in epithelia. In addition to the epithelial keratins, sometimes called cytokeratins, there are the keratins that form hair and nail, the so-called hard keratins. These proteins share similar structural properties and are distinct from the β-keratins of avian feather. The keratins have been divided into two types: the acidic or type I keratins and the neutral basic or type II keratins. To form a filament requires a 1:1 ratio of type I to type II keratins (see Steinert and Parry, 1985). It appears that almost any combination of type I and type II keratins will form a filament, although there is evidence that the composition of a keratin filament affects its structure (Eichner et al., 1986). Keratins copolymerize only with other keratins and not with other IFP types. When vimentin and keratin are expressed in the same cell, they form discrete filament systems (Aubin et al., 1980; Henderson and Weber, 1981).

The type IV IFPs, i.e., α-internexin, the neurofilament proteins NFL, NFM, NFH, and nestin, resemble the type III IFPs in their polymerization properties. *In vitro* NFI, NFM and α-internexin are capable of homopolymeric assembly (Balin and Lee, 1991). *In vivo* type IV IFPs copolymerize with vimentin (Fliegner and Liem, 1991). α-internexin is capable of forming an extended IF system *in vivo* (Ching and Liem, 1993), whereas NFI, N-FM, and NFH appear to be unable to form an IF network on their own, but must copolymerize with some other type III or type IV IF protein (Ching and Liem, 1993; Lee et al., 1993). Type III and type IV IFPs can also be distinguished by the structure of their tail domains and the pattern of introns and exons in the genes encoding them. Type IV IFP genes have an intron/exon pattern completely different from all other vertebrate and invertebrate IFP genes (Fliegner and Liem,

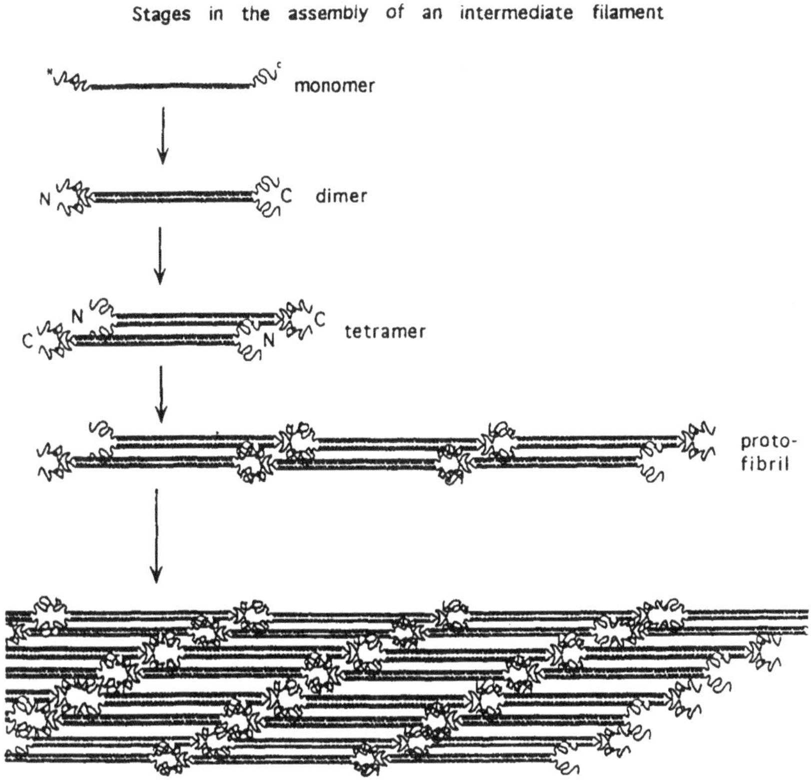

Figure 2. ***The structure of an IF.*** Here we have illustrated the steps in the assembly of an IF. The monomeric subunit proteins dimerize in a parallel orientation. These dimers in turn dimerize in an antiparallel and staggered manner to form a tetramer. Tetramers assemble end to end to form protofilaments; between six to eight protofilaments align laterally to form the final filament. (This scheme is based on the model proposed by Steinert et al., 1993).

1991). Lewis and Cowan (1986) originally suggested that the type IV IFP genes originated in the reverse transcription of an mRNA (Figure 1).

INTERMEDIATE FILAMENT STRUCTURE

IFs are hierarchically organized (Figure 2) (see Parry and Steinert, 1992; Steinert, et al., 1993). IFP monomers form a coiled-coil dimer in which the two monomers are in parallel and in register with one another. In the case of the keratins, dimer is formed from a type I and a type II keratin protein. Next, two dimers combine in an antiparallel and staggered manner to form a tetramer (Potschka et al., 1990; Geisler et al., 1992; Steinert et al., 1993). Tetramers associate with one another in an end-to-end fashion to form a protofilament; protofilaments associate laterally to form the final IF (Figure 2). IFs isolated from the cell are heterogeneous and appear to contain between six to eight protofilaments (Aebi et al., 1988).

Both microfilaments and microtubules are polar polymers, i.e., each end is chemically unique. The polarity of these polymers is fundamental to their function as tracks along which mechanoenzymes transport cargo. In contrast, the antiparallel organization of the IFP tetramer generates a polymer without an intrinsic polarity, and so it is unable to support similar directed motions.

The structure of IFs provides them with unique mechanical properties *vis-à-vis* the other cytoskeletal systems (Figure 3) (Janmey et al., 1991; Ingber 1993). When force is applied to an IF, the IF first stretches and eventually stiffens; however, even at quite high stress levels it does not break. When the force is removed, the IF relaxes back to its original length. Microtubules and microfilaments, in contrast, are quite brittle and break at relatively low stress levels. The molecular nature of IF elasticity remains to be completely understood. It could be due to the deformation of the subunit proteins themselves, or to a force-induced reorganization of IF structure.

IFs are highly resistant to salt extraction and generally require severe denaturants, e.g., 8-9M urea or 6M guanidine HCl, to be solubilized. In the cell, the measured half life of IFPs is on the order of 15-30 hours (McTavish et al. 1983; Coleman and Lazarides, 1992). These data lead to the idea that IFs are static structures. In fact, it is now clear that IFs are quite dynamic. However, IF dynamics differ from those of microtubules and microfilaments. First, IF assembly is a simple equilibrium process that does not involve nucleotide hydrolysis (Zackroff and Goldman, 1979). Second, while microtubules and microfilaments grow and shrink primarily by means of subunit addition (or release) from the polymer ends, IF assembly occurs at multiple sites along the length of the filament (Ngai et al., 1990). The nature of these sites is not yet clear. There does not appear to be a specific nucleation site for IF assembly in the cell. When IF assembly is initiated *de novo,* assembly starts at multiple sites scattered throughout the cytoplasm (Kreis et al., 1983; Magin et al., 1990; Raats et al., 1990; Sarria et al., 1990).

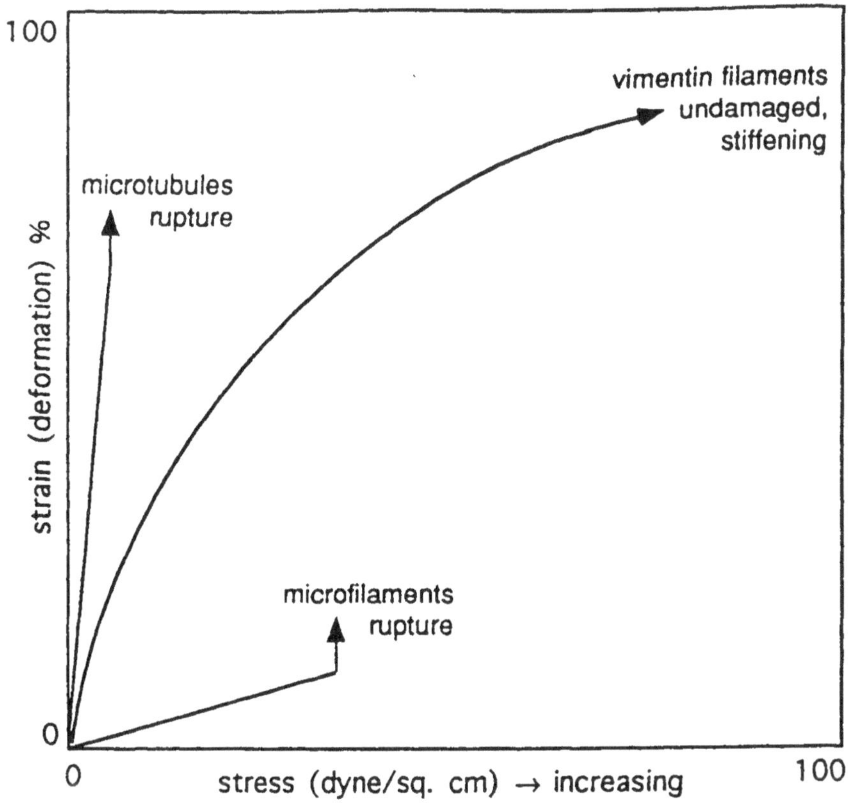

Figure 3. Mechanical properties of IFs. This graph is derived from the work of Jammey et al. (1991) and illustrates the behavior of vimentin-type IFs under mechanical stress and in comparison to the behavior of microtubules and microfilaments. Microtubules are very stiff and break at very low stress levels. Microfilaments are slightly stronger, but also break at relatively low stress levels. In contrast, vimentin filaments deform as stress is applied, stretching to almost twice their original length. At higher stress levels, vimentin filaments become stiff, but do not break.

When cells are treated with detergent, between 0.5 to 5% of the total IFP present appears in a soluble, tetrameric form. The tetrameric form of IFPs is thought to be the smallest form to be observed in living cells (Soellner et al., 1985), and appears to be the basic IF assembly intermediate, much as the (α/β tubulin dimer and the actin monomer are the basic assembly intermediates of microtubules and microfilaments, respectively. In the case of the keratins, there is evidence that a discrete soluble, higher order oligomer may play an important role in assembly (Bachant, 1993; Bachant & Klymkowsky, submitted).

How IFP tetramer and higher order oligomers form is unclear. At present there is no evidence for the involvement of chaperoning proteins (Bachant, 1993; Bachant & Klymkowsky, submitted).

IF ORGANIZATION

IFs lack the ability for supported directed motility and there is no specific IF-organizing center (see above). Therefore, interactions between IFs and other components appear necessary to generate an organized IF system within the cell. IF organization is cell-type specific, suggesting that different sets of interactions are involved in different cell types. For example, vimentin usually forms a radial system in most cells, but in vascular endothelial cells it forms a ring structure around the nucleus (Blose and Meltzer, 1981). Upon conversion from fibroblastic to adipocytic morphology, vimentin filaments change from a radial system and come to surround the lipid droplets (Franke et al., 1987). Similarly, the IFs of myoblasts are organized in a largely radial system, but upon muscle maturation most IFs come to be closely associated with either the periphery of the Z-discs or the sarcolemma (Lazarides et al., 1981; Holtzer et al., 1982). These differences in IF organization appear to be mediated by cell-type specific factors. Studies in the *Xenopus* oocyte (Dent et al., 1992; Bachkus and Klymkowsky, unpublished observations) indicate that there are factors that can actively suppress the ability of IFPs to form extended IF networks.

In most cells in which they are expressed, type III IFPs are organized primarily through their interactions with microtubules and microfilaments (Figure 3). The microtubule motor protein kinesin appears to drag IFs toward the periphery of the cell (Gyoeva and Gelfand, 1991). At the periphery, IFs appear to interact with microfilaments, which act to move IFs toward the cell center (Hollenbeck et al., 1989; Tint et al., 1991). In the normal cell, there is a balance between these two opposing forces. It is a common observation that a wide range of conditions, including heat shock, viral infection, and many toxic compounds lead to the collapse of IF organization (see Klymkowsky, 1988; Klymkowsky et al., 1988; Doedens et al., 1994). It seems likely that these affect IF organization by altering the balance between microfilament $\leftrightarrow$ IF $\leftrightarrow$ microtubule interactions. The disruption of IF organization seen in

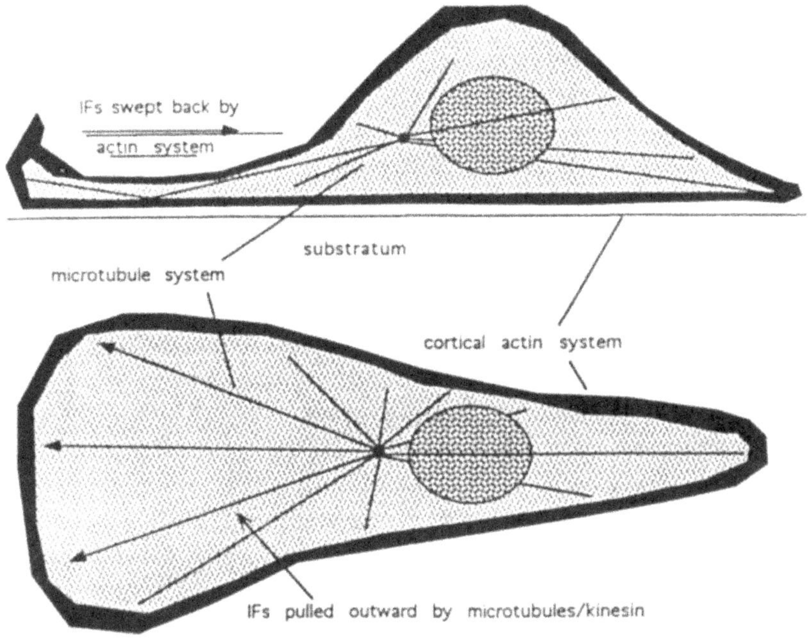

Figure 4. MTs, MFs, and IF organization. In a typical fibroblastic cell, IFs interact with both microtubules and microfilaments. Interactions with microtubules drag the IFs toward the periphery of the cell. This movement appears to be mediated by the plus end microtubule motor protein kinesin (see text). At the periphery of the cell, IFs interact with the cortical microfilament system. This interaction sweeps the IFs back toward the cell center. In a normal cell, IF organization is determined in large measure by the balance between these two interactions.

many pathological states (see below) may be due to a perturbation of this balance. Aberrant accumulations of IFs could, in turn, lead to further cellular dysfunction. (See Lee and Cleveland, 1994)

A number of experiments indicate that IFs are basically a passive element in the cytoskeleton. Disruption of IF organization does not affect either microtubule or microfilament organization in a gross way (see Klymkowsky et al., 1983, 1989 and references therein), although subtle changes may have gone unnoticed. IFs can clearly interact with one another, however (Klymkowsky, 1982). For example, in epidermal cells keratin filaments are often found laterally aligned in large bundles. These "tonofilaments" appear to provide increased mechanical strength (see below). Whether this lateral alignment is mediated by IF-associated proteins or by direct interactions between IFPs remains unclear. In neurons, the large tail domains of the neurofilament proteins NFM and NFH project laterally from the filament axis. These projections block the close apposition of filaments and are involved in the even spacing of neurofilaments within the axon. The presence of such a neurofilament lattice plays a major role in maintaining the caliber of large axons (see below).

IFs also interact with more specialized cellular structures. For example, in epithelia and cardiac muscle, IF bundles are associated with specialized cellular adherence junctions, desmosomes and hemidesmosomes. Junctions ultrastructurally similar to desmosomes have been found in neural cells and in the eye lens (see Schwarz et al., 1990). In striated muscle (i.e., cardiac and skeletal), IFs are closely associated with the periphery of the Z-disc (see Tokuyasu et al. 1984; Tokuyasu et al., 1985 and references therein) In skeletal muscle, IFs are concentrated at the myotendinous junction, the site at which the thin (actin) filaments of muscle are anchored (Tidball, 1992).

At the desmosome Ca^{2+}-dependent cellular adhesion molecules, or cadherins interact to form the cell-cell attachments (see Koch et al., 1992). Their tail domains appear to nucleate the formation of the cytoplasmic portion of the desmosome (see Troyanovsky et al., 1993). Keratin, vimentin, and desmin-type IFs have been observed to attach laterally to this cytoplasmic material. While the exact molecular details of this interaction are not known, interactions between a cytoplasmic component of the desmosome, desmoplakin, and IFs have been observed *in vivo* (Stappenbeck and Green, 1992; Kouklis et al., 1994).

PHOSPHORYLATION AND IF ORGANIZATION

In cells IFPs are present in both phosphorylated and nonphosphorylated forms (Gard et al., 1979). The rate of phosphate turnover is far more rapid than the turnover rate of the IFP (McTavish et al., 1983) suggesting that phosphorylation is involved in the regulation of IF organization. IFP types characteristically differ in their molar phosphate content (Steinert et al., 1982). Vimentin filaments appear to be the least highly substituted (~1 mole phosphate/mole protein) (Steinert et al., 1982), while neurofilaments are the most highly substituted (9-22 moles phosphate/mole protein) (Jones and Williams, 1982;

Ksiezak-Reding and Yen, 1987). IFP phosphorylation sites are located in both the N-terminal head and C-terminal tail domains (Geisler et al., 1987; Evans, 1988; Lee et al., 1988; Steinert, 1988; Shea et al., 1990; Sihag and Nixon, 1991).

Interest in IFP phosphorylation has been stimulated by the observation that the phosphorylation of IFPs increases during M-phase of the cell cycle (Bravo et al., 1982; Evans and Fink, 1982; Celis et al., 1983; Westwood et al., 1985; Zieve, 1985; Tolle et al., 1987; Chou et al., 1989, 1990; Rosevear et al., 1990; Skalli et al., 1992). IFP phosphorylation *in vitro* can induce the disassembly of IFs (Inagaki et al., 1987, 1988; Kitamura et al., 1989; Hisagana et al., 1990; Yano et al., 1991). Given that there is often a dramatic reorganization of IFs upon entry into mitosis (Aubin et al., 1980; Zieve et al., 1980; Horwitz et al., 1981; Blose and Bushnell, 1982; Franke et al., 1982; Lane et al., 1982; Franke et al., 1984; Tumer and Ruane, 1985; Rosevear et al., 1990), it has been speculated that phosphorylation is the direct cause of the M-phase reorganization of IFs. The most dramatic example of M-phase reorganization of IFs occurs during the meiotic maturation of the *Xenopus* oocyte, in which keratin-type IFs are disassembled into soluble oligomers (Klymkowsky et al., 1987, 1991). In a particularly elegant series of experiments, Nishizawa et al. (1991) showed that the phosphorylation of a specific serine residue in GFAP occurs only in the region of the contractile ring and leads to the local disassembly of IFs. Such local disassembly of IFs presumably facilitates the separation of daughter cells during cytokinesis.

Considerable efforts have been made to identify the protein kinase activities involved in IFP phosphorylation. IFPs are well-characterized substrates for cAMP dependent protein kinase (O'Connor et al., 1981; Gard and Lazarides, 1982; Ando et al., 1989; Geisler et al., 1989; Eckert and Yeagle, 1990; Sihag and Nixon, 1991), protein kinase C (Inagaki et al., 1988; Sihag et al. 1988; Ando et al., 1989; Geisler et al., 1989; Kitamura et al., 1989; Yano et al., 1991), calmodulin-dependent kinase (Ando et al., 1989; Yano et al., 1991) and the cell cycle regulating cdc2 kinase (MPF kinase) (Chou et al., 1990, 1991; Guan et al., 1992; Kusubata et al., 1993). In fact, isoforms of protein kinase C have been found to be physically associated with IFPs (Murti et al., 1992; Omary et al., 1992; Spudich et al., 1992). In specific cultured cell lines IFP phosphorylation has been shown to be increased by treatment with isoproterenol (Gard and Lazarides, 1982), norepinephrine (Browing and Sanders, 1981), FSH (Spruill et al., 1983), epidermal growth factor

(Baribault et al., 1989), and angiotensin II (Tsuda et al., 1988). In general, agents which raise cytoplasmic [Ca^{2+}] or cAMP appear to influence the state of IF phosphorylation, although potential roles of other second messengers remain to be determined. Taken together, these studies indicate that IFP phosphorylation is a complex process, involving type-specific, multi-site phosphorylation by multiple protein kinases.

In neurons, phosphorylation of the neurofilament proteins occurs predominantly in the axons and has been linked with the lateral projection of the tail domains (Hisanaga and Hirokawa, 1989). A distinct neurofilament kinase, which might mediate this phosphorylation, has been described (Julien et al., 1983; Wible et al., 1989; Hollander and Bennett, 1992). Dephosphorylation of neurofilament proteins may be involved in the closer packing of neurofilaments at the nodes of Ranvier (Mata et al., 1992). The role of interphase IFP phosphorylation in non-neuronal cells is less well understood. Soellner et al. (1985) reported that the small soluble pool of tetrameric vimentin in mammalian cells does not appear to be more highly phosphorylated than the protein in insoluble assembled filaments. Similar results have been found for keratin IFs (Bachant, 1993). Lamb et al. (1989) microinjected purified cAMP-dependent protein kinase, which induces the disassembly of vimentin filaments *in vitro*, into cultured cells. They observed an increase in vimentin phosphorylation, but no apparent disassembly of vimentin filaments.

Altered IFP phosphorylation and changes in IF organization have been associated with some pathological conditions. For example, elevated neurofilament phosphorylation accompanies the characteristic changes in neurofilament organization observed in amyotrophic lateral sclerosis (ALS) (Manetto et al., 1988; Matsumoto et al., 1992), aluminum toxicity (Bizzi and Gambetti, 1986), and Lewy body formation in Parkinson's disease (Fomo et al., 1986). Desmin hyperphosphorylation has been found in some familial myopathies (Rappaport et al., 1988) and there is evidence that elevated phosphorylation may be associated with the pathological accumulation of keratins in Mallory bodies (Cadrin et al., 1992). It is not clear whether increased phosphorylated IFP is a cause or an effect of these pathological changes.

CHANGES IN IF ORGANIZATION ASSOCIATED WITH HUMAN DISEASES

Changes in IF organization have been linked to a number of human disease states. The rare genetic disorder Giant Axonal Neuropathy is characterized by accumulations of neurofilaments in the distal portions of neurons and in a number of other cell types (see Klymkowsky et al., 1988); similar changes in the neuronal IF organization occur following hexacarbon solvent and acrylamide intoxication (Davenport et al., 1976). Epidermolysis bullosum simplex (EBS) is identified in part by the presence of keratin aggregates in the basal cell layer of the epidermis (see below). Cytoplasmic inclusions of keratin (Mallory bodies) found in hepatocytes of patients with alcoholic liver disease, also called alcoholic hyaline, were first described by Mallory (1911) (see Worman, 1990). Mallory bodies are most commonly associated with chronic ethanolism, but have also been observed in other degenerative liver conditions and some hepatomas. A dramatic accumulation of keratin filaments has also been associated with Cushing's syndrome. Originally described in 1935 as Crooke's hyaline change (Crooke, 1935), this involves accumulation of densely packed keratin-type IFs in the ACTH secreting cells of the pituitary (Dedicco et al., 1972; Robert et al., 1978; Neumann et al., 1984). In addition, some disorders associated with changes in IF organization may actually involve IF-associated proteins. For example, ichthyosis vulgaris has been linked to mutations in the IF-associated protein filaggrin (Sybert et al., 1985; Kanitakis et al., 1988). Filaggrin is required for the normal maturation of keratinized epithelium, and is capable of cross-linking keratin filaments *in vitro* (Dale et al., 1978). Whether the normal role of filaggrin *in vivo* is to cross-link keratin filaments remains controversial (Weidenthaler et al., 1993).

Antibodies directed against neurofilament epitopes recognize the filament tangles found in ALS (Itoh et al., 1992), Matsumoto et al., 1992; Troost et al., 1992) (see below), in the Lewy bodies found in Parkinson's disease (Hill et al., 1991; Pollanen et al., 1992), and in other neurodegenerative disorders (reviewed in Pollanen et al., 1993). A recent report suggested that peripherin type IFs may also be affected (Migheli et al., 1993).

Myopathies characterized by accumulation of intrasarcoplasmic granular and filamentous, desmin-containing material have been described (see Goebel and Bornemann, 1993). Among these rare disorders are distal myopathies (Edstrom et al., 1980; Goebel et al., 1980), myopathies with cardiomyopathy (Fardeau et al., 1978), or cardiomyopathies (Sakakibara et al., 1970; Porte et al., 1980). Most of the reported cases appear to be hereditary, autosomal dominant disorders although one family with an X-linked myopathy has also been studied (Pette and Staron, 1990) and a few sporadic cases have also been reported (Shafiq et al., 1974).

In the mammalian retina, the most abundant glial cell type is termed Muller cell. While Muller cells are believed to be functionally equivalent to astrocytes (Ripps and Witkovsky, 1985), they normally express vimentin rather than GFAP (Bignami and Dahl, 1979). In response to hereditary retinal degeneration, detachment, or following light-induced injury, Muller cells begin to express GFAP (Bignami and Dahl, 1979; Eisenfeld et al., 1984; Sarthy and Fu, 1989). The significance of this change in IF gene expression with injury is not known.

CELL TYPE-SPECIFIC EXPRESSION OF IFPS IN VERTEBRATES

Different IFP types are expressed in different cell types. During development, the simple epithelial keratins K8, K18, and K19 are the first IFPs to appear. Keratin-like IFs have also been identified in the oocytes of invertebrate species (Boyle and Ernst, 1989; Schroeder and Otto, 1991). As embryonic development proceeds, the pattern of IFP expression changes. Vimentin is the first non-keratin IFP to appear and it is initially found in embryonic mesoderm, i.e., in nonepithelial, non-endothelial cells. Finally, the tissue-specific IFPs desmin, GFAP, peripherin, nestin, α-internexin, NFI, NFM, and NFH are expressed. In epithelial tissues the simple epithelial keratins are replaced by other keratins. For example, in the epidermis, keratins K5 and K14 replace K8/KI8 in the basal layer of the epidermis; as the epidermis stratifies, K5/K14 are in turn replaced by K1/K1O in the suprabasal layers. During transitional periods, a cell can express two types of IFP simultaneously. In the developing mammalian nervous system, cells first express keratin, then vimentin, and finally either neuronal IFPs or GFAP (Cochard and Paulin, 1984; Kamada et al., 1988; Oudega and Marani, 1991). In the adult, IFP expression is generally stable. Even following neoplastic transformation, cells will usually continue to express the IFP of their original state. This permits tumors of ambiguous aspect to be classified as carcinomas (keratin-expressing), sarcomas (vimentin-expressing), etc. (see below).

IFP gene expression is similar to that of other genes. Both positive and negative regulatory regions have been identified (see Sax et al., 1988, 1990; Zehner, 1991; Desmarais et al., 1992; Zehner, et al., 1992; Neznanov and Oshima, 1993 as examples). Moreover, some of these regulatory regions appear to be located downstream of the transcription start site (Neznanov and Oshima, 1993). In the invertebrates, the differential splicing of IFP transcripts is quite frequent (see Way et al., 1992;

Dodemont et al., 1994). In vertebrates, vimentin (Capetanaki et al., 1983; Zehner and Paterson, 1983a,b) and peripherin (Landon et al., 1989) have been reported to be differentially spliced. In both cases, the difference in splicing does not affect the final protein product. Following translation, IFPs appear to be remarkably stable proteins, with half-lives on the order of 20 hours and greater (Coleman and Lazarides, 1992; Bachant, 1993). The one dramatic exception to this rule occurs when a type I or type II keratin is synthesized in the absence of an appropriate pairing partner; unpaired keratins have a rather short half-life (Knapp et al., 1989; Lu and Lane, 1990).

CELLULAR FUNCTIONS OF IFs AND HUMAN DISEASE

Understanding the functions of IFPs has lagged behind that of microtubules and microfilaments for two reasons. First, the reagents required to disrupt IF organization are rather "high tech," i.e., anti-IFP antibodies (see Klymkowsky et al., 1989), dominantly mutated forms of IFPs, "knock out" (Baribault and Oshima, 1991; Baribault et al., 1993) or spontaneously occurring mutations in IFP genes. In contrast, there are naturally occurring, small molecular weight compounds that can be used to disassemble or stabilize microfilaments and microtubules. Second, manipulation of IF network organization has thus far produced only very subtle phenotypes in cultured cells (see Klymkowsky et al., 1989). At this point only a few clearly demonstrated *in vitro* phenotypes have been reported. There is the striking correlation between vimentin expression and cholesterol esterification (Sarria et al., 1992). The association between vimentin filaments and viral components appears to be important for efficient replication of frog virus 3 (Murti et al., 1988). The expression of a vimentin cDNA has been reported to suppress the chemically-transformed phenotype of BHK SN-10 cells (Eiden et al., 1991), and there is a correlation between the expression of the simple epithelial keratin K8/Kl8 and cellular invasiveness/malignancy in melanoma cells (see Ben-Ze'ev et al., 1986; Chu et al., 1993). In each of these cases, however, the effects are subtle, requiring very specific assays to reveal their presence.

In this light, many investigators have turned to the intact organism to study IF function. In *Xenopus* the use of oligonucleotides (Torpey et al., 1992) and injected anti-keratin antibodies (Klymkowsky et al., 1992) indicates that the presence of an intact keratin filament system is essential for normal gastrulation. Anti-IF antibody injected experiments have also been performed on the early mouse embryo (Emerson, 1988) where it was found that the disruption of keratin filament organization had no obvious effect on the ability of the embryo to form a trophectoderm or an apparently normal blastocyst. These studies have been extended by Baribault and Oshima (1991) and Baribault et al. (1993) who constructed transgenic mice lacking a functional K8 keratin gene. K8 is the first type II keratin to be expressed in the mouse embryo and is the only type II keratin present in many embryonic tissues. In its absence, keratin filaments cannot form. Such K8-minus mice pass through the early stages (i.e., up to stage 12) of mouse development apparently normally, but show a very high level of embryonic lethality (~95%) between embryonic days 12 and 13. Internal bleeding is apparent and there are gross morphological changes in the liver. Apparently keratin filaments are critical for normal liver organization/function, although their exact role remains unclear. There are also defects in placental organization and K8 minus female mice are sterile (Baribault et al., 1993). Moreover, the phenotype of the K8 knock out mutation is significantly affected by genetic background. In certain genetic backgrounds the frequency of embryonic survival is greatly increased, while in the adult a new phenotype, namely extensive hyperplasia of the colon and large intenstine appears (Baribault et al., 1994). The mechanism by which the absence of an intact keratin filament network leads to such a hyperplasia is not clear.

Two other IFP genes have been mutated in a similar manner, vimentin (Colucci et al., 1994) and GFAP (Gomi et al., 1995; Pekny et al., 1995). Here the phenotypes are much more subtle and superficially at least the mutant mice behave quite normally.

KERATIN FILAMENTS AND THE EPIDERMIS

As hemoglobin is to blood cells, so keratins are to the epidermis. It is therefore not surprising that mutations in keratins should give rise to epidermal dysfunction. The demonstration of this fact is one of the most powerful examples of the use of molecular genetic technology in resolving the molecular basis of disease. Albers and Fuchs (1987) found that truncated forms of the human K14 type I keratin protein would disrupt the organization of preexisting keratin filaments when expressed in cultured cells. Transgenic mice in which such truncated keratins were expressed in the basal cell layer of the epidermis displayed a skin defect similar to that seen in the autosomal dominant human disease epidermolysis bullosum simplex (EBS) (Vassar et al., 1991): namely, they show severe skin blistering upon mechanical insult. At approximately the same time there were a number of reports that EBS was associated with mutations in the K5 (Lane et al., 1992) or K14 (Bonifas et al., 1991; Coulombe et al., 1991; Compton et al., 1992; Humphries et al., 1993) genes. The mutated keratin was, in some cases, found to disrupt keratin filament organization when expressed in cultured epidermal cells (Coulombe et al., 1991). Subsequently, mutations in the suprabasal keratins Kl and K10 have been linked to another human skin disease, epidermolytic hyperkeratosis (EHK) (Cheng et al., 1992; Chipev et al., 1992; Fuchs et al., 1992; Rothnagel et al., 1992). A growing number of mutations have now been identified in EBS/EHK patients. The most severe, in terms of effects on keratin filament organization, correspond to the Dowling- Meara type of EBS, the least severe with Wayne-Cockayne (see Fuchs and Coulombe, 1992). Mutations in other epidermal keratin are linked to other skin diseases (see McLean & Lane, 1995).

Ultrastructural analysis of transgenic mice expressing dominant mutated forms of K14, or patients with EBS, indicate that basal cells lyse in a specific region, between the basement membrane and the nucleus (Figure 5). That this cellular defect is the direct result of the mechanical weakening of these cells following the disruption of their keratin filament network is suggested by the observation of similar phenotype in people with the equivalent of a null mutation in an epidermal keratin gene (see McLean & Lane, 1995).

STUDIES ON NEUROFILAMENT FUNCTION

It has been assumed that neurofilaments play a role in the determination of axonal caliber (i.e., diameter) (see Muma et al., 1991). Dramatic evidence in support of this hypothesis comes from two independent observations. The first involves the naturally occurring *Quiver* (Yamasaki et al., 1991) and hypotrophic axonopathy h(ax) (Mizutani et al., 1992) mutants of the Japanese quail. These animals were identified on the basis of a behavioral phenotype, namely a mild quiver of the head and body.

Ultrastructural examination reveals that the myelinated axons of these animals are significantly smaller in diameter than those of wild type animals and are devoid of neurofilaments. Neurofilaments were also absent from the cell bodies (Yamasaki et al., 1991). Molecular analysis indicates that these defects are due to a nonsense mutation in the gene for NFL (Ohara et al., 1993).

Transgenic mice that express a mutated form of the human NFH protein in their neurons have also been constructed (Eyer and Peterson,

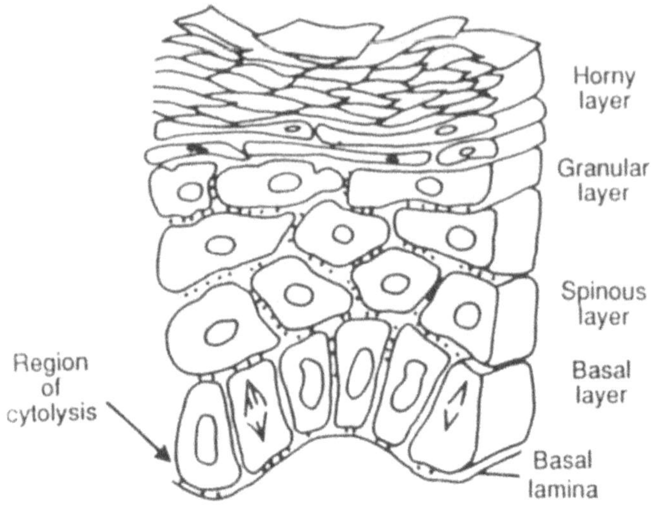

Figure 5. Cellular site of epidermal keratin-based dysfunction. The epiderm
express the keratin K5 and K14. In the spinous layer keratins K1 and K10 are expr
these keratins lead to blistering of the skin in response to mechanical insult. It is th
leads to the fragility of the cells. In the case of mutations in K5/K15, cellular frag
(From Klymkowsky, 1991).

1994). The presence of the mutated protein induces the collapse of the cell's neurofilament system and the formation of huge bundles of intact neurofilaments in the neuronal cell body. In these mice essentially no neurofilaments are present in the axons and the diameter of the largest axon is reduced to about half of the value seen in wild type animals. In contrast to the *Quv* mutant quail, these mice behave quite normally (Eyer, and Peterson, 1994). Together these studies indicate that neurofilaments are required to maintain the caliber of neurons. Whether this is their only function remains to be determined.

DESMIN FILAMENTS AND MUSCLE

Schultheiss et al. (1991) transfected cultured chick muscle cells with a dominant negative mutation of the IFP desmin. The mutant protein disrupted the organization of the endogenous IF system, but had no apparent effect on muscle cell differentiation or the formation of the contractile apparatus. Since the association of desmin with the Z-disc is a relatively late event in muscle differentiation (Hill et al., 1986), this result was, to some extent, expected. The limitation of this study is that cultured muscle is not under the same kind of mechanical load as is muscle *in situ*. Much as mechanical insult is required to reveal the epidermal defects caused by keratin mutations (see above), so the disruption of desmin filament organization requires mechanical activity to reveal an effect on muscle structure. To study the role of desmin in skeletal muscle, we have introduced mutated forms of desmin into the dorsal myotomal muscle of the *Xenopus* tadpole. We find that disruption of the desmin filament network produces specific defects in the attachment of myofibrils to the intersomite junction (Cary & Klymkowsky, 1995). The intersomite junction of the *Xenopus* myotomal muscle is the functional and structural equivalent of the myotendinous junction of adult skeletal muscle. The concentration of desmin at the intersomite junction (Cary & Klymkowsky, 1994a), the myotendinous junction (Tidball, 1992), and the intercalated discs of cardiac muscle (Kartenbeck et al., 1983; Thornell et al., 1984; Samuel et al., 1985; Osinka & Lemanski, 1989), together with the phenotype observed following the disruption of desmin organization (Cary & Klymkowsky, 1995), suggests that in muscle, as in the epidermis, IFs are involved in reinforcing cellular structures routinely subject to physical stress. Based on these data, we predict that defects in desmin will be found to be a contributing factor for both myopathies and cardiomyopathies.

ECTOPIC EXPRESSION OF IFPs

There are a number of studies that have examined the effects of ectopic or supraphysiological expression of IFPS. For example, in transgenic mice, ectopic expression of desmin (Dunia et al., 1990) or increased expression of chick vimentin (Capetanaki et al., 1989) in the lens leads to cataracts. Similarly, ectopic expression of an epidermal keratin Kl in the pancreas of transgenic mice induces a diabetes-type syndrome, presumably due to damage of islet cells (Blessing et al., 1993). Overexpression of neurofilament proteins (Côté et al. 1993; Xu et al., 1993) produces symptoms similar to those of ALS in transgenic mice. Finally, over-expression of desmin or vimentin in the myocytes of the Xenopus tadpole leads to defects in cellular morphology and apparent cell death (Cary and Klymkowsky, 1994b). These studies indicate that the simple over- or inappropriate synthesis of IFPs can produce severe cellular pathologies.

IF TYPING AND TUMORS

The general observation that IFPs can be used as markers for the histotypic origin of cells has become an important aid in the differential diagnosis of human tumors (Duboulay, 1985; Dar et al., 1992; Leong, 1992). Many tumors can be classified as epithelial, mesenchymal, glial, neuronal, or muscle in origin based on the IFPs present. This is of special value in cases where tumors lack readily identifiable or distinguishing histological features. In general, carcinomas will express keratins, nonmuscle sarcomas, lymphomas, leukemias: melanomas will express vimentin: muscle sarcomas will contain desmin: gliomas and glial derived tumors will contain GFAP, and neuronally derived tumors will primarily express neurofilament proteins (Osborn & Weber, 1983). The use of IFP typing has been of importance in identifying small round cell tumors in children, particularly in the diagnosis of rhabdomyosarcoma (Molenaar et al., 1985; Eusebi et al. 1986; Osborn et al., 1986; Parham et al., 1991; Kodet et al., 1993), and to subclassify carcinomas such as adenocarcinoma and squamous cell carcinoma (Osborn et al., 1985), based on the spectrum of individual keratin proteins that are expressed (Moll et al., 1982, 1983, 1992; Blobel et al. 1984, 1985; Denk et al. 1985; Lifschitzmercer et al., 1991; Smedts et al., 1993).

In certain types of tumors, however, IFPs are inappropriately expressed or ambiguous. Tumors containing both vimentin and keratin include: mesothelioma (Larocca and Rheinwald, 1984; Churg, 1985;

Mayall et al., 1992), nephroblastoma (Altmannsberger et al., 1984; Denk et al., 1985), renal cell carcinoma (Holthofer et al., 1983; Dierick et al. 1991), synovial sarcoma (Salisbury and Isaacson, 1985), some small cell carcinomas of the lung (Gatter et al., 1986), some prostatic carcinomas (Bussemakers et al., 1992; Mcmahon et al., 1992), thyroid carcinoma (Livolsi, 1992; Yamamoto et al., 1992), and metastatic tumors present in ascites and pleural fluids (Ramaekers et al., 1983). Perhaps the most surprising finding has been the demonstration of both vimentin and keratin immunostaining in a significant number of tumors of smooth muscle origin (Brown et al., 1987; Norton et al. 1987). A critical question, then, is how reliable is IFP typing. It may well be that the degree of differentiation within the tumor and its relation to the sequence of IFP expression during normal development must be carefully considered, particularly if the normal cell counterpart contains multiple IFPs. During embryonic development, cells of parietal endoderm (Lane et al., 1983; Lehtonen et al., 1983), kidney (Holthofer et al., 1984), prostate (Wernert et al., 1987), and neural tube (Bennett, 1987) apparently undergo this sort of transition. Since many tumors appear to be caricatures of normal differentiation and express fetal or embryonic antigens (Knapp and Franke, 1989), it is possible that the coexpression of vimentin and keratin in some renal and prostatic carcinomas (Herman et al., 1983; Holthofer et al., 1983; Nazeer et al., 1991), reflects the differentiated state of these tumors. Finally, Knapp and Franke (1989) suggested that transcriptionally inactive IF genes may be intrinsically unstable and give rise to spontaneous IFP expression upon oncogenic transformation. There are also tumors, e.g., Burkitt's lymphomas, plasmacytoma and myeloma cells that do not express any IFPs (Dellagi et al., 1984; Lilienbaum et al., 1986; Paulin-Levasseur et al., 1988).

IFPs AS PROGNOSTIC INDICATORS FOR TUMOR TREATMENT

Recent studies have indicated that inappropriate IFP expression in tumors may have prognostic value. A good example comes from studies of breast cancers, where choices for treatment are increasingly influenced by prognostic indicators. Normal mammary epithelium contains keratin (Gould et al., 1990; Taylor-Papadimitriou et al., 1992). A number of studies, however, indicate that the acquisition of vimentin in high grade ductal carcinomas of the breast is a marker for a poor prognosis (Domagala et al. 1990; Bocker et al., 1992; Sommers et al. 1992; Taylor-Papadimitriou et al., 1992; Guelstein et al., 1993). The appearance of vimentin is reported to be related to increased invasiveness of these tumors (Thompson et al., 1991, 1992). Although not as extensively studied, a similar relationship between the inappropriate acquisition of vimentin expression and increased tumor aggressiveness has also been suggested for adrenal carcinomas (Henzen-Logmans et al., 1988; Miettinen, 1992). In the case of melanomas, the result is different. These cells normally express vimentin, but where keratins are coexpressed, malignancy is found to be significantly increased (Ben-Ze'ev et al., 1986; Fuchs et al., 1992; Hendrix et al., 1992; Klymkowsky, 1995).

SUMMARY

In summary, it is now clear that mutations in IFPs and defects in IF organization are likely to be the cause, or significant contributors, to a range of pathological conditions. By turning to animal models, the role of IFs in cells and disease is likely to become much more clear in the next few years. In any case, immunohistochemical typing of IFs has become a useful tool for the surgical pathologist, provided that a clear understanding of the complexity of IFP expression is kept in mind.

ACKNOWLEDGMENTS

Our work has been supported by grants from National Science Foundation, the American Cancer Society, and the Colorado Chapter of the American Heart Association (M.W.K.) and the National Institutes of Health (R.M.E.).

REFERENCES

Aebi, U., Cohn, J., Buhle, L., & Gerace, L. (1986). The nuclear lamina is a meshwork of intermediate-type filaments. *Nature* 323; 560-564.

Aebi, U., Haner, M., Troncoso, J., Eichner, R., & Engel, A. (1988). Unifying principles in intermediate filament (IF) structure and assembly. Protoplasma 145, 73-81.

Albers, K. & Fuchs, E. (1987). The expression of mutant epidermal keratin cDNAs transfected in simple epithelial and squamous cell carcinoma lines. *J. Cell Biol.* 105, 791-806.

Albers, K. & Fuchs, E. (1992). The molecular biology of intermediate filament proteins. *Int. Rev. Cytol.* 134, 243-279.

Altmannsberger, M., Osborn, M., Schafer, H., Schauer, A., & Weber, K. (1984). Distinction of nephroblastomas from other childhood tumors using antibodies to intermediate filaments. Virchows Arch. (Cell Pathol) 45, 113-124.

Ando, S., Tanabe, K., Gonda, Y., Sato, C., & Inagaki, M. (1989). Domain and sequence-specific phosphorylation of vimentin induces disassembly of filament structure. Biochemistry 28, 2974-2979.

Aubin, J. E., Osborn, M., Franke, W. W., & Weber, K. (1980). Intermediate filaments of the vimentin-type and the cytokeratin-type are distributed differently during mitosis. Exp. Cell Res 129, 149-165.

Bachant, J. B. (1993). Thesis. University of Colorado, Boulder.

Balin, B. J. & Lee, V. M. Y. (1991). Individual neurofilament subunits reassembled in vitro exhibit unique biochemical, morphological and immunological properties. Brain Res, 556, 196-208.

Baribault, H., Blouin, R., Bourgon, L., & Marceau, N. (1989). Epidermal growth factor-induced selective phosphorylation of cultured rat hepatocyte 55-kD cytokeratin before filament reorganization and DNA synthesis. J Cell Biol. 109, 1665-1676.

Baribault, H. & Oshima, R. G. (1991). Polarized and functional epithelia can form after the targeted inactivation of both mouse keratin 8 alleles. J. Cell Biol, 115, 1675-1684.

Baribault, H., Penner, J. Iozzo, R.V. & Wilson, M.H. (1994). Colorectal hyperplasia and inflammation in keratin 8-deficient FVB/N mice. Genes Dev. 8, 2964-2973.

Baribault, H., Price, J., Miyai, K., & Oshima, R. G. (1993). Mid-gestational lethality in mice lacking keratin 8. Genes Devel. 7, 1191-1202.

Bartnik, E. & Weber, K. (1989). Widespread occurrence of intermediate filaments in invertebrates: Common principles and diversion. Eur. J. Cell Bio, 50, 17-33.

Ben-Ze'ev, A., Zoller, M., & Raz, A. (1986). Differential expression of intermediate filament proteins in metastatic and nonmetastatic variants of the BSp73 tumor. Cancer Res. 46, 785-790.

Bennett, G. S. (1987). Changes in intermediate filament composition during neurogenesis. Curr. Top Dev. Biol. 21, 151-183.

Bershadsky, A. D. & Vasiliev, J. M. (1988). *Cytoskeleton.* Plenum. New York & London.

Bignami, A. & Dahl, D. (1979). The radial glia of Muller in the rat retina and their response to injury. An immunofluorescence study with antibodies to the glial fibrillary acidic (GFA) protein. Exp. Eye Res, 28, 62-69.

Bizzi, A. & Gambetti, P. (1986). Phosphorylation of neurofilaments is altered in aluminum intoxication. Acta Neuropathol, 71, 154-158.

Blessing, M., Ruther, U., & Franke, W. W. (1993). Ectopic synthesis of epidermal cytokeratins in pancreatic islet cells of transgenic mice interferes with cytoskeletal order and insulin production. J. Cell Biol. 120, 743-755.

Blobel, G. A., Moll, R., Franke, W. W. Kayser, K. W., & Gould, V. E. (1985). The intermediate filament cytoskeleton of malignant mesotheliomas and its diagnostic significance. Am. J. Pathol, 121, 235-247.

Blobel, G. A., Moll, R., Franke, W. W., & Vogt-Moykopf, I. (1984). Cytokeratins in normal lung and lung carcinomas. 1. Adenocarcinomas, squamous cell carcinomas and cultured cell lines. Virchows Arch. (Cell Pathol.) 45, 407-429.

Blose, S. & Bushnell, A. (1982). Observations on the vimentin-10-nm filaments during mitosis in BHK21 cells. Exp. Cell Res. 142, 57-62.

Blose, S. H. & Meltzer, D. 1. (1981). Visualization of the 10-nm filament vimentin rings in vascular endothelial cells in situ: close resemblance to vimentin cytoskeletons found in monolayers in vitro. Exp. Cell Res. 135, 299-309.

Bocker, W., Bier, B., Freytag, G., Brommelkamp, B., Jarasch, E. D., Edel, G., Dockhomdwomiczak, B. & Schmid, K. W. (1992). An immunohistochemical study of the breast using antibodies to basal and luminal keratins, a-smooth muscle actin, vimentin, collagen-IV and laminin 2 epitheliosis and ductal carcinoma in situ. Virchows Arch. [A] 421, 323-330.

Bonifas, J. M., Rothman, A. L. & Epstein, E. H., Jr. (1991). Epidermolysis bullosa simplex: evidence in two families for keratin gene abnormalities. Science 254, 1202-1205.

Boyle, J. A. & Emst, S. G. (1989). Sea urchin oocytes possess cortical arrays of microfilaments, microtubules and intermediate filaments. Devel. Biol. 134, 72-84.

Bravo, R., Small, J. V., Fey, S. J., Mose-Larsen, P., & Celis, J. E. (1982). Architecture and polypeptide composition of HeLa cytoskeletons. Modification of cytoarchitectural polypeptides during mitosis. J. Mol. Biol. 154, 121-143.

Bray, D. (1992). Cell Movements. Garland Publishing, New York & London.

Browing, E. T. & Sanders, M. M. (1981). Vimentin: a phosphoprotein under hormonal control. J. Cell Biol. 90, 803-808.

Brown, D. C., Theaker, J. M., Banks, P. M., Gatter, K. C., & Mason, D. Y. (1987). Cytokeratin expression in smooth muscle and smooth muscle tumors. Histopathology 11, 477-486.

Bussemakers, M. J. G., Verhaegh, G. W. C. T., Vanbokhoven, A., Debruyne, F. M. J. & Schalken, J. A. (1992). Differential expression of vimentin in rat prostatic tumors. Biochem. Biophys. Res. Commun. 182, 1254-1259.

Cadrin, M., Mcfarlaneanderson, N., Aasheim, L. H., Kawahara, H., Franks, D. J., Marceau, N. & French, S. W. (1992). Differential phosphorylation of CK8 and CK18 by 12-o- tetradecanoyl-phorbol-13-acetate in primary cultures of mouse hepatocytes. Cell Signal. 4, 715-722.

Capetanaki, Y., Smith, S., & Heath, J. P. (1989). Overexpression of the vimentin gene in transgenic mice inhibits normal lens cell differentiation. J. Cell Biol 109, 1653-1664.

Capetanaki, Y. G., Ngai, J., Flytzanis, C. N., & Lazarides, E. (1983). Tissue-specific expression of two MRNA species transcribed from a single vimentin gene. Cell 35, 411-420.

Cary, R. B. & Klymkowsky, M. W. (1994a). Desmin organization during the differentiation of the dorsal myotome in *Xenopus laevis*. *Differentiation*, in press.

Cary, R.B. & Klymkowsky, M.W. (1994b). Differential organization of desmin and vimentin in muscle is due to differences in their head domains. J Cell Biol., 126, 445-456.

Cary, R.B. & Klymkowsky, M.W. (1995). Disruption of intermediate filament organization leads to structural defects at the intersomite junction in *Xenopus* myotomal muscle. Development 122, 1041-1052.

Celis, J. E., Mose-Larsen, P., Fey, S. J., & Celis, A. (1983). Phosphorylation of keratin and vimentin polypeptides in normal and transformed mitotic human epithelial amnion cells: behavior of keratin and vimentin filaments during mitosis. J. Cell Biol. 97, 1429-1434.

Cheng, J., Syder, A. J., Yu, Q. C., Letai, A., Paller, A. S., & Fuchs, E. (1992). The genetic basis of epidermolytic hyperkeratosis: a disorder of differentiation-specific epidermal keratin genes. Cell 70, 811-819.

Ching, G.Y. & Liem, R.H. (1993). Assembly of type IV neuronal intermediate filaments in nonneuronal cells in the absence of preexisting cytoplasmic intermediate filaments. J. Cell Biol. 122, 1323-1335.

Chipev, C. C., Korge, B. P., Markova, N., Bale, S. J., DiGiovanna, J. J., Compton, J. G., & Steinert, P. M. (1992). A leucine → proline mutation in the HI subdomain of keratin I causes epidermolytic hyperkeratosis. Cell 70, 821-82.

Choi, B. H. & Kim, R. C. (1984). Expression of glial fibrillary acidic protein in immature oligodendroglia. Science 223, 407-409.

Chou, Y. H., Bischoff, J. R., Beach, D. & Goldman, R. D. (1990). Intermediate filament reorganization during mitosis is mediated by p34cdc2 phosphorylation of vimentin. Cell 62, 1063-1071.

Chou, Y. H., Ngai, K. L., & Goldman, R. (1991). The regulation of intermediate filament reorganization in mitosis—p34cdc2 phosphorylates vimentin at a unique N-terminal site. J. Biol. Chem. 266, 7325-7328.

Chou, Y. H., Rosevear, E., & Goldman, R. D. (1989). Phosphorylation and disassembly of intermediate filaments in mitotic cells (published erratum appears in Proc. Natl. Acad. Sci. USA, 1989, 86, 32141). Proc. Natl. Acad. Sci. USA 86, 1885-1889.

Chu, Y. W., Runyan, R. B., Oshima, R. G., & Hendrix, M. J. (1993). Expression of complete keratin filaments in mouse L cells augments cell migration and invasion. Proc. Natl. Acad. Sci. USA 90, 4261-4265.

Churg, A. (1985). Immunohistochemical staining for vimentin and keratin in malignant mesothelioma. *Am.* J. Surg. Pathol. 9, 360-365.

Cochard, P. & Paulin, D. (1984). Initial expression of neurofilaments and vimentin in the central and peripheral nervous system of the mouse embryo in vivo. J. Neurosci 4, 2080- 2094.

Cohen, C. & Parry, D. A. D. (1986). α-helical coiled-coils—a widespread motif in proteins. TIBS 11, 245-248.

Coleman, T. R. & Lazarides, E. (1992). Continuous growth of vimentin filaments in mouse fibroblasts. J. Cell Sci. 103, 689-698.

Colucci, E., Portier, M.M., Dunia, I., Paulin, D., Pournin, S. & Babinet, C. (1994). Mice lacking vimentin develop and reproduce without an obvious phenotype. Cell 79, 679-694.

Compton, J. G., DiGiovanna, J. J., Santucci, S. K., Keams, K. S., Amos, C. I., Abangan, D. L., Korge, B. P., McBride, 0. W., Steinert, P. M., & Bale, S. J. (1992). Linkage of epidermolytic hyperkeratosis to the type 11 keratin gene cluster on chromosome 12q. Nat. Genet. 1(4), 301-305.

Côté, F., Collard, J.-F., & Julien, J.-P. (1993). Progressive neuronopathy in transgenic mice expressing the human neurofilament heavy gene: a mouse model of amyotrophic lateral sclerosis. Cell 73, 35-46.

Coulombe, P. A., Hutton, M. E., Letal, A., Hebert, A., Paller, A. S., & Fuchs, E. (1991). Point mutations in human keratin 14 genes of epidermolysis bullosa simplex patients: Genetic and functional analyses. Cell 66, 1301-1311.

Coulombe, P. A., Hutton, M. E., Vassar, R., & Fuchs, E. (1991). A function for keratins and a common thread among different types of epidermolysis bullosa simplex diseases. J. Cell Biol. 115, 1661-1674.

Crick, F. H. C. (1953). The fourier transform of a coiled-coil. *Acta Cryst.* 6, 685-688.

Crooke, A. C. (1935). A change in the basophil cells of the pituitary gland common to conditions which exhibit the syndrome attributed to basophil adenoma. *J.* Pathol. Bacteriol. 41, 339-349.

Dale, B. A., Holbrook, K. A. & Steinert, P. M. (1978). Assembly of stratum comeum basic protein and keratin filaments in microfibrils. Nature 276, 729-731.

Dar, A. U. H., Hird, P. M., Wagner, B. E., & Underwood, J. C. E. (1992). Relative usefulness of electron microscopy and immunocytochemistry in tumour diagnosis—10 years of retrospective analysis. J. Clin. Pathol. 45, 693-696.

Davenport, J. G., Farrell, D. F., & Sumi, S. M. (1976). 'Giant axonal neuropathy' caused by industrial chemicals. Neurology 26, 919-923.

Dedicco, F. A., Dekker, A. & Yunis, E. J. (1972). Fine structure of Crooke's hyaline change in the human pituitary gland. Arch. Pathol. 94, 65-70.

Dellagi, K., Broue, J. C., Portie, M. M., & Lenoir, G. M. (1984). Abnormal expression of vimentin intermediate filaments in human lymphoid cell lines with deletion or translocation of the distal end of chromosome 8. J. Natl. Cancer Inst. 73, 95-99.

Dellagi, K., Vainchenker, W., Vinci, G., Paulin, D., & Brouet, J. (1983). Alteration of vimentin intermediate filaments during differentiation of human hemopoietic cells. EMBO J. 2, 1509-1511.

Denk, H., Weybora, W., Ralschek, M., Sohar, R., & Franke, W. W. (1985). Distribution of vimentin, cytokeratins, and desmosomal plaque proteins in human nephroblastoma as revealed by specific antibodies: co-existence of cell groups of different degrees of epithelial differentiation. Differentiation 29, 88-97.

Dent, J. A., Cary, R. B., Bachant, J. B., Domingo, A. & Klymkowsky, M. W. (1992). Host cell factors controlling vimentin organization in the *Xenopus* oocyte. *J. Cell Biol.* 119, 855-866.

Desmarais, D., Filion, M., Lapointe, L. & Royal, A. (1992). Cell-specific transcription of the peripherin gene in neuronal cell lines involves a cis-acting element surrounding the TATA box. EMBO J. 11, 2971-2980.

Dierick, A. M., Praet, M., Roels, H., Verbeeck, P., Robyns, C. & Oosterlinck, W. (1991). Vimentin expression of renal cell carcinoma in relation to DNA content and histological grading—a combined light microscopic, immunocytochemical and cytophotometrical analysis. Histopathology 18, 315-322.

Dodemont, H., Riemer, D., Ledger, N., & Weber K (1994). Eight genes and alternative RNA processing pathways generate an unexpectedly large diversity of cytoplasmic intermediate filament proteins in the nematode *Caenorhabditis elegans*. EMBO J. 13, 2625-2638.

Doedens, J., Maynell, L. A., Klymkowsky, M. W., & Kirkegaard, K. (1994). Secretory pathway function, but not cytoskeletal integrity, is required in poliovirus infection. In: New Aspects of Positive-strand RNA Viruses (Brinton, M.A. & Heinz, F.X., Eds.), pp. 159-172, Amer. Society Microbiology, Washington, D.C.

Domagala, W., Lasota, J., Dukowicz, A., Markiewski, M., Striker, G., Weber, K., & Osborn, M. (1990). Vimentin expression appears to be associated with poor prognosis in node-negative ductal NOS breast carcinomas. Am. J. Pathol. 137, 1299-1304.

Duboulay, C. E. H. (1985). Immunocytochemistry of soft tissue tumors: a review. J. Pathol 146, 77-94.

Dunia, I., Pieper, F., Manenti, S. van de Kemp, A., Devilliers, G. Benedetti, E. L. & Bloemendal, H. (1990). Plasma membrane-cytoskeleton damage in eye lenses of transgenic mice expressing desmin. Eur. J. Cell Biol. 53, 59-74

Eckert, B. S. & Yeagle, P. L. (1990). Modulation of keratin intermediate filament distribution in vivo by induced changes in cyclic AMP-dependent phosphorylation. Cell Motil. Cytoskel. 17, 291-300.

Edstrom, L., Thomell, L., & Eriksson, A. (1980). A new type of hereditary distal myopathy with characteristic sarcoplasmic bodies and intermediate (skeletin) filaments. J. Neurol. Sci. 47, 171-190.

Eichner, R., Sun, T.-T., & Aebi, U. (1986). The role of keratin subfamilies and keratin pairs in the formation of human epidermal intermediate filaments. J. Cell Biol. 102, 1767-1777.

Eiden, M., MacArthur, L. & Okayama, H. (1991). Suppression of the chemically transformed phenotype of BHK cells by a human cDNA. Mol. Cell. Biol. 11, 5321-5329.

Eisenfeld, A. J., Bunt-Milam, A. & Sarthy, P. V. (1984). Muller cell expression of glial acidic fibrillary protein after genetic and experimental photoreceptor degeneration in the rat retina. Invest. Ophthal. Vis. Sci. 25, 1321-1328.

Emerson, J. A. (1988). Disruption of the cytokeratin filament network in the preimplantation mouse embryo. Development 104, 219-234.

Eusebi, V., Ceccarelli, C., Gorza, L., Schiaffino, S. & Bussolati, G. (1986). Immunochemistry of rhabdomyosarcoma: the use of four different markers. *Am.* J. Surg. Pathol. 10, 293-299.

Evans, R. M. (1988). The intermediate-filament proteins vimentin and desmin are phosphorylated in specific domains. Eur. J. Cell Biol. 46, 152-160.

Evans, R. M. & Fink, L. M. (1982). An alteration in the phosphorylation of vimentin-type intermediate filaments is associated with mitosis in cultured mammalian cells. Cell 29, 43-52.

Eyer, J. & Peterson, A. (1994). Neurofilament-deficient axons and perikaryal aggregates in viable transgenic mice expressing a neurofilament-beta-galactosidase fusion protein. Neuron 12, 389-405.

Fardeau, M., Godet-Guillain, J., Tome, F. M. S., Collin, H., Gaudeau, S., Boffety, C., & Vemant, P. (1978). Une nouvelle affection musculaire familiale definie par l'accumulation intrasarcoplasmique d'un material granulofilamentaire dense en microscopic electronique. Rev. Neurol. (Paris) 134, 411-425.

Fliegner, K. H. & Liem, R. K. (1991). Cellular and molecular biology of neuronal intermediate filaments. Intl. Rev. Cytol. 131, 109-167.

Fomo, L. S., Stemberger, L. A., Stemberger, N. H., Strefling, A. M., Swanson, K. & Eng, L. F. (1986). Reaction of Lewy bodies with antibodies to phosphorylated and nonphosphorylated neurofilaments. Neurosci. Lett. 64, 253-258.

Franke, W. W., Grund, C., Kuhn, C., Lehto, V., & Virtanen, 1. (1984). Transient change of organization of vimentin filaments during mitosis as demonstrated by a monoclonal antibody. Exp. Cell Res. 154, 567-580.

Franke, W. W., Hergt, M., & Grund, C. (1987). Rearrangement of the vimentin cytoskeleton during adipose conversion: formation of an intermediate filament cage around lipid globules. Cell 49, 131-141.

Franke, W. W., Schmid, E., Grund, C., & Geiger, B. (1982). Intermediate filament proteins in nonfilamentous structures: transient disintegration and inclusion of subunit proteins in granular aggregates. Cell 30, 103-113.

Fuchs, E. & Coulombe, P. A. (1992). Of mice and men—genetic skin diseases of keratin. Cell 69, 899-902.

Fuchs, E., Esteves, R. A., & Coulombe, P. A. (1992). Transgenic mice expressing a mutant keratin 10 gene reveal the likely genetic basis for epidermolytic hyperkeratosis. Proc. Natl. Acad Sci. USA 89, 6906-6910.

Fuchs, U., Kivela, T., Summanen, P., Immonen, I., & Tarkkanen, A. (1992). An immunohistochemical and prognostic analysis of cytokeratin expression in malignant uveal melanoma. Am. J. Pathol, 141, 169-181.

Gard, D. L., Bell, P. B., & Lazarides, E. (1979). Coexistence of desmin and the fibroblastic intermediate filament subunit in muscle and nonmuscle cells. *Proc, Natl. Acad, Sci. USA* 76, 3894-3898.

Gard, D. L. & Lazarides, E. (1982). Cyclic AMP-modulated phosphorylation of intermediate filament proteins in cultured avian myogenic cells. Mol. Cell. Biol. 2, 1104-1114.

Gatter, K. C., Dunnill, M. S., van Muijen, G. N. P., & Mason, D. Y. (1986). Human lung tumors may co-express different classes of intermediate filaments. J. Clin. Pathol. 39, 950-954.

Geisler, N., Hatzfeld, N., & Weber, K. (1989). Phosphorylation of vimentin by protein kinase A and C is restricted to the head domain: identification of the phosphoserine sites and their influence on filament formation. Eur. J. Biochem. 183, 441-447.

Geisler, N., Vandekerckhove, J., & Weber, K. (1987). Location and sequence characterization of the major phosphorylation sites of the high molecular mass neurofilament proteins M and H. FEBS Lett. 221, 403-407.

Geisler, N., Schunemann, J., & Weber, K. (1992). Chemical cross-linking indicates a staggered and antiparallel protofilament of desmin intermediate filaments and characterizes one higher-level complex between protofilaments. Eur. J. Biochem. 206, 841-852.

Gerace, L. & Burke, B. (1988). Functional organization of the nuclear envelope. Ann, Rev. Cell Biol. 4, 335-374.

Giese, G. & Traub, P. (1986). Induction of vimentin synthesis in mouse myeloma cells MEPC- 11 by 12-0-tetradecanoylphorbol-13- acetate. Eur. J. Cell Biol. 40, 266-274.

Goebel, H. & Bornemann, A. (1993). Desmin pathology in neuromuscular diseases. *Virchows Arch.* (Cell Pathol). 64, 127-135.

Goebel, H. H., Lenard, H., Langenbeck, U., & Mehl, B. (1980). A form of congenital muscular dystrophy. Brain Dev. 2, 337-400.

Gomi, H., Yokoyama, T., Fujimoto, K., Ikeda, T., Katoh, A., Itoh, T., & Itohara, S. (1995). Mice devoid of the glial fibrillary acidic protein develop normally and are susceptible to scrapie prions. Neuron 14, 29-41.

Gould, V. E., Koukoulis, G. K., Jansson, D. S., Nagle, R. B., Franke, W. W., & Moll, R. (1990). Coexpression patterns of vimentin and glial filament protein with cytokeratins in the normal, hypesplastic, and neoplastic breast. Am. J. Pathol. 137, 1143-1155.

Gounari, F., Merdes, A., Quinlan, R., Hess, J., FitzGerald, P. G., Ouzounis, C. A. & Georgatos, S. D. (1993). Bovine filensin possesses primary and secondary structure similarity to intermediate filament proteins. *J Cell Biol.* 121, 847-853.

Guan, R. J., Hall, F. L., & Cohlberg, J. A. (1992). Proline-directed protein kinase (p34cdc2/p58cyclin A) phosphorylates bovine neurofilaments. J. Neurochem. 58, 1365-1371.

Guelstein, V. I., Tchypysheva, T. A., Ermilova, V. D., & Ljubimov, A. V. (1993). Myoepithelial and basement membrane antigens in benign and malignant human breast tumors. Intl. J. Cancer 53, 269-277.

Gyoeva, F. K. & Gelfand, V. I. (1991). Coalignment of vimentin intermediate filaments with microtubules depends on kinesin. Nature 353, 445-448.

Hedberg, K. K. & Chen, L. B. (1986). Absence of intermediate filaments in a human adrenal conex carcinoma derived cell line. Exp. Cell Res. 163, 509-517.

Henderson, D. & Weber, K. (1981). Immunoelectron microscopical identification of the two types of intermediate filaments in established epithelial cells. Exp. Cell Res. 132, 297-311.

Hendrix, M. J. C., Seftor, E. A., Chu, Y. W., Seftor, R. E. B., Nagle, R. B., McDaniel, K. M., Leong, S. P. L., Yohem, K. H., Leibovitz, A. M., Meyskens, F. L., Conaway, D. H., Welch, D. R., Liotta, L. A. & Stetler stevenson, W. (1992). Coexpression of vimentin and keratins by human melanoma tumor cells—correlation with invasive and metastatic potential. J. Natl. Cancer Inst. 84, 165-174.

Henzen-Logmans, S. C., Stel, H. V., van Muijen, G. N. P., Mullink, H., & Meijer, C. J. L. M. (1988). Expression of intermediate filament proteins in adrenal cortex and related tumors. Histopathology 12, 359-372.

Herinan, C. J., Moesker, O., Kant, A., Huysmans, A., Vooijs, G. P., & Ramaekers, F. C. S. (1983). Is renal cell (Grawitz) tumor a carcinosarcoma? Evidence from analysis of intermediate filament types. Virchows Arch, (Cell Pathol.) 44, 73-83.

Hill, C. S., Duran, S., Lin, Z. X., Weber, K., & Holtzer, H. (1986). Titin and myosin, but not desmin, are linked during myofibrillogenesis in postmitotic mononucleated myoblasts. J. Cell. Biol. 103, 2185-2196.

Hill, W. D., Lee, V. M. Y., Hurtig, H. I., Murray, J. M. & Trojanowski, J. Q. (1991). Epitopes located in spatially separate domains of each neurofilament subunit are present in Parkinson's disease Lewy bodies. J. Comp. Neurol. 309, 150-160.

Hisagana, S., Gonda, Y., Inagaki, M., Ikai, A. & Hirokawa, N. (1990). Effect of phosphorylation of the neurofilament L protein on filamentous structures. Cell Regul. 1, 237-248.

Hisanaga, S. & Hirokawa, N. (1989). The effects of dephosphorylation on the structure of the projections of neurofilament. J. Neurosci. 9, 959-966.

Hollander, B. A. & Bennett, G. S. (1992). Characterization of a neurofilament-associated kinase that phosphorylates the middle molecular mass component of chicken neurofilaments. Brain Res. 599, 237-245.

Hollenbeck, P. J., Bershadsky, A. D., Pletjushkina, 0. Y., Tint, 1. S. & Vasiliev, J. M. (1989). Intermediate filament collapse is an ATP-dependent and actin-dependent process. J. Cell Sci. 92, 621-631.

Holthofer, H., Miettinen, A., Lehto, V., Lehtonen, E., & Virtanen, I. (1984). Expression of vimentin and cytokeratin types of intermediate filament proteins in developing and adult human kidneys. Lab. Invest. 50, 552-559.

Holthofer, H., Miettinen, A., Paasivuo, M. A., Lehto, V., Linder, E., Alfthan, O., & Virtanen, 1. (1983). Cellular origin and differentiation of renal cell carcinomas: a fluorescence microscopy study with kidney-specific antibodies, anti-intermediate filament antibodies, and lectins. Lab. Invest. 49, 317-326.

Holtzer, H., Bennett, G. S., Tapscott, S. J., Croop, J. M., & Toyama, Y. (1982). Intermediate-sized filaments: changes in synthesis and distribution in cells of the myogenic and neurogenic lineages. Cold Spring Harb. Symp. Quant, Biol. 42, 317-329.

Horwitz, B., Kupfer, H., Eshhar, Z. & Geiger, B. (1981). Reorganization of arrays of prekeratin filaments during mitosis: Immunofluorescence microscopy with multiclonal and monoclonal prekeratin antibodies. *Exp. Cell Res.* 134, 281-290.

Humphries, M. M., Sheils, D. M., Faffar, G. J., Kumar, S. R., Kenna, P. F., Mansergh, F. C., Jordan, S. A., Young, M., & Humphries, P. (1993). A mutation (Met→Arg) in the type I keratin (K14) gene responsible for autosomal dominant epidermolysis bullosa simplex. Hum. Mutat. 2, 37-42.

Inagaki, M., Gonda, Y., Matsuyama, M., Nishizawa, K., Nish, Y. I. & Sato, C. (1988). Intermediate filament reconstitution in-vitro. The role of phosphorylation on the assembly-disassembly of desmin. J. Biol. Chem. 263, 5970-5978.

Inagaki, M., Nishi, Y., Nishizawa, K., Matsuyama, M., & Sato, C. (1987). Site-specific phosphorylation induces disassembly of vimentin filaments in vitro. Nature 328, 649-652.

Ingber, D. E. (1993). Cellular tensegrity - defining new rules of biological design that govern the cytoskeleton. J. Cell Sci. 104, 613-627.

Itoh, T., Sobue, G., Ken, E., Mitsuma, T., Takahashi, A., & Trojanowski, J. Q. (1992). Phosphorylated high molecular weight neurofilament protein in the peripheral motor, sensory and sympathetic neuronal perikarya—system-dependent normal variations and changes in amyotrophic lateral sclerosis and multiple system atrophy. Acta Neuropathol. 83, 240-245.

Janmey, P. A., Euteneuer, U., Traub, P., & Schliwa, M. (1991). Viscoelastic properties of vimentin compared with other filamentous biopolymer networks. J. Cell Biol. 113, 155-160.

Jones, S. M. & Williams, R. C. (1982). Phosphate content of mammalian neurofilaments. J. Biol. Chem. 257, 9902-9905.

Julien, J. P., Smoluk, G. D., & Mushynski, W. E. (1983). Characteristics of the protein kinase activity associated with rat neurofilament preparations. Biochim. Biophys. Acta 755, 25-31.

Kamada, H., Itoh, T., Hyogo, T., Satoh, S., Ogasawara, T., Fujiwara, H., Ara, S., Hotta, T., Suematsu, K., Nakamura, J., & et. al. (1988). [Immunohistochemical study of developing rat embryo—localization of vimentin, GFAP, neurofilament protein within rat embryo central nervous system]. No To Shinkei 40, 211-218.

Kanitakis, J., Ramirez, A., Reano, A., Viac, J., Roche, R., & Thivolet, J. (1988). Filaggrin expression in normal and pathological skin: a marker of keratinocyte differentiation. Virchows Arch. [*A*]. 412, 375-382.

Karnovsky, A. & Klymkowsky, M.W. (1995). Anterior axis duplication in *Xenopus* induced by the over-expression of the cadherin-binding proteins plakoglobin. Proc. Natl. Acad. Sci. USA., in press.

Kartenbeck, J., Franke, W. W., Moser, J. G., & Stoffels, U. (1983). Specific attachment of desmin filaments to desmosomal plaques in cardiac myocytes. EMBO J. 2, 735-742.

Kitamura, S., Ando, S., Shibata, M., Tanabe, K., Sato, C., & Inagaki, M. (1989). Protein kinase C phosphorylation of desmin at four serine residues within the non-helical head domain. J. Biol. Chem. 264, 5674-5678.

Klymkowsky, M.W. (1982). Vimentin and keratin intermediate filament systems in cultured PtK2 epithelial cells are interrelated. EMBO J. 2, 161-165.

Klymkowsky, M. W. (1988). Metabolic inhibitors and intermediate filament organization in human fibroblasts. Exp. Cell Res. 174, 282-290.

Klymkowsky, M.W. (1991). Intermediate filaments: getting under the skin. Nature 354, 264-265.

Klymkowsky, M.W. & Karnovsky, A. (1994). Morphogenesis and the cytoskeleton: Studies of the Xenopus embryo. Devel. Biol. 126, 372-384.

Klymkowsky, M. W., Miller, R. H. & Lane, E. B. (1983). Morphology, behavior, and interaction of cultured epithelial cells after the antibody-induced disruption of keratin filament organization. J. Cell Biol. 96, 494-509.

Klymkowsky, M. W., Bachant, J. B. & Domingo, A. (1989). Functions of intermediate filaments. Cell Motil. Cytoskel. 14, 309-331.

Klymkowsky, M. W., Christian, E., Ham, R. G., Plummer, D. J. & Sprenger, F. (1988). Giant axonal neuropathy mutations, intermediate filaments, and cellular metabolism. Alan R. Liss, New York,

Klymkowsky, M. W., Maynell, L. A. & Nislow, C. (1991). Cytokeratin phosphorylation, cytokeratin filament severing and the solubilization of the maternal MRNA Vgl. J. Cell Biol. 114, 787-797.

Klymkowsky, M. W., Maynell, L. A., & Polson, A. G. (1987). Polar asymmetry in the organization of the cortical cytokeratin system of Xenopus laevis oocytes and embryos. Development 100, 543-557.

Klymkowsky, M. W., Shook, D. R. & Maynell, L. A. (1992). Evidence that the deep keratin filament system of the Xenopus embryo acts to ensure normal gastrulation. Proc. Natl. Acad. Sci. USA 89, 8726-8740.

Klymkowsky, M.W. (1995). Intermediate filaments: new proteins, some answers, more questions. Curr. Op. Cell Biol. 7, 46-54.

Knapp, A. C., Bosch, F. X., Hergt, M., Kuhn, C., Winter, S. S., Schmid, E., Regauer, S., Bartek, J. & Franke, W. W. (1989). Cytokeratins and cytokeratin filaments in subpopulations of cultured human and rodent cells of nonepithelial origin: modes and pattems of formation. Differentiation 42, 81-102.

Knapp, A. C. & Franke, W. W. (1989). Spontaneous losses of control of cytokeratin gene expression in transformed, non-epithelial human cells occurring at different levels of regulation. Cell 59, 67-79.

Koch, P. J., Goldschmidt, M. D., Zimbelmann, R., Troyanovsky, R., & Franke, W. W. (1992). Complexity and expression pattems of the desmosomal cadherins. Proc. Natl. Acad. Sci. USA. 89, 353-357.

Kodet, R., Newton, W. A., Hamoudi, A. B., Asmar, L., Jacobs, D. L. & Maurer, H. M. (1993). Childhood rhabdomyosarcoma with anaplastic (pleomorphic) features—a report of the intergroup rhabdomyosarcoma study. Am. J. Surg. Pathol. 17, 443-453.

Kouklis, P. D., Hutton, E., & Fuchs, E. (1994). Making a connecion: direct binding between keratin intermediate filaments and desmosomal proteins. J. Cell Biol. 127, 1049-1060.

Kreis, T. E., Geiger, B., Schmid, E., Jorcano, J. L. & Franke, W. W. (1983). De novo synthesis and specific assembly of keratin filaments in nonepithelial cells after microinjection of mRNA for epidermal keratin. Cell 32, 1125-1137.

Ksiezak-Reding, H. & Yen, S. (1987). Phosphatase and carbocyanine dye binding define different types of phosphate groups in mammalian neurofilaments. J. Neurosci. 7, 3554-3560.

Kusubata, M., Matsuoka, Y., Tsujimura, K., Ito, H., Ando, S., Kamijo, M., Yasuda, H., Ohba, Y., Okumura, E., Kishimoto, T. & Inagaki, M. (1993). cdc2 kinase phosphorylation of desmin at 3 serine/threonine residues in the amino-terminal head domain. Biochem, Biophys. Res. Commun. 190, 927-934.

Lamb, N. J. C., Fernandez, A., Feramisco, J. R., & Welch, W. J. (1989). Modulation of vimentin containing intermediate filament distribution and phosphorylation in living fibroblasts by the cAMP-dependent protein kinase. J. Cell Biol. 108, 2409-2422.

Landon, F., Lemonnier, M., Benarous, R., Huc, C., Fiszman, M., Gros, F. & Portier, M. M. (1989). Multiple mRNAs encode peripherin, a neuronal intermediate filament protein. Embo J. 8, 1719-1726.

Lane, E. B., Goodman, S. L., & Trejdosiewicz, L. K. (1982). Disruption of the keratin filament network during epithelial cell division. EMBO J. 1, 1365-1372.

Lane, E. B., Hogan, B. L., Kurkinen, M., & Carrels, J. 1. (1983). Co-expression of vimentin and cytokeratins in parietal endoderm cells of early mouse embryo. Nature 303, 701-704.

Lane, E. B., Rugg, E. L., Navsaria, H., Leigh, I. M., Heagerty, A. H., Ishida, Y. A., & Eady, R. A. (1992). A mutation in the conserved helix termination peptide of keratin 5 in hereditary skin blistering. Nature 356, 244-246.

Larocca, P. J. & Rheinwald, J. G. (1984). Coexpression of simple epithelial keratins and vimentin by human mesothelium and mesothelioma in-vivo and in culture. Cancer Res. 44, 2991-2999.

Lazarides, E., Granger, B. L., Gard, D. L., O'Connor, C. M., Breckler, J., Price, M., & Danto, S. I. (1981). Desmin- and vimentin-containing filaments and their role in the assembly of the Z disk in muscle cells. Cold Spring Harbor Symp. Quant. Biol. 46, 351-378.

Lee, M.K. & Cleveland, D.W., (1994) Neurofilament function and dysfunction: involvement in axonal growth and neuronal disease. Curr. Opin. Cell Biol. 6, 34-40.

Lee, V. M., Otvos, L., Carden, M. J., Hollosi, M., Dietzschold, B., & Lazzarini, R. A. (1988). Identification of the major multiphosphorylation site in mammalian neurofilaments. Proc. Natl. Acad. Sci. USA 85, 1998-2002.

Lee, M. K., Xu, Z., Wong, P. C., & Cleveland, D. W. (1993). Neurofilaments are obligate heteropolymers in vivo. J. Cell Biol. 122, 1337-1350.

Lehtonen, E., Lehto, V. P., Vartio, T., Badley, R. A. & Virtanen, 1. (1983). Expression of cytokeratin polypeptides in mouse oocytes and preimplantation embryos. Dev. Biol. 100, 158-165.

Leong, A. S. Y. (1992). New vistas in the histopathological assessment of cancer. Med. J. Aust. 157, 699-701.

Lewis, S. A. & Cowan, N. J. (1986). Anomalous placement of introns in a member of the intermediate filament multigene family: an evolutionary conundrum. *M*ol. Cell Biol. 6, 1529-1534.

Lifschitzmercer, B., Fogel, M., Moll, R., Jacob, N., Kushnir, I., Livoff, A., Waldheff, R., Franke, W. W. & Czemobilsky, B. (1991). Intermediate filament protein profiles of human testicular non-seminomatous germ cell tumors—correlation of cytokeratin synthesis to cell differentiation. Differentiation. 48, 191-198.

Lilienbaum, A., Legagneux, V., Portier, M., Dellagi, K., & Paulin, D. (1986). Vimentin gene: expression in human lymphocytes and in Burkitt's lymphoma cells. EMBO J. 5, 2809- 2814.

Livolsi, V. A. (1992). Papillary neoplasms of the thyroid—pathologic and prognostic features. Am. J. Clin. Pathol. 97, 426-434.

Lu, X. & Lane, E. (1990). Retrovirus-mediated transgenic keratin expression in cultured fibroblasts: specific domain functions in keratin stabilization and filament formation. Cell 62, 681-696.

Magin, T. M., Bader, B. L., Freudenmann, M. & Franke, W. W. (1990). De novo formation of cytokeratin filaments in calf lens cells and cytoplasts after transfection with cDNAs or microinjection with mRNAs encoding human cytokeratins. Eur J. Cell. Biol. 53, 333-348.

Mallory, F. B. (1911). Cirrhosis of the liver: five different types of lesions from which it may arise. Bull. Johns Hopkins Hosp. 2. 22, 69-75.

Manetto, V., Stemberger, N. H., Perry, G., Stemberger, L. A. & Gambetti, P. (1988). Phosphorylation of neurofilaments is altered in amyotrophic lateral sclerosis. J. Neuropath. Exp. Neurol. 47, 642-653.

Mata, M., Kupina, N. & Fink, D. J. (1992). Phosphorylation-dependent neurofilament epitopes are reduced at the node of Ranvier. J. Neurocytol. 21, 199-210.

Matsumoto, S., Kusaka, H., Murakami, N., Hashizume, Y., Okazak, H., & Hirano, A. (1992). Basophilic inclusions in sporadic juvenile amyotrophic lateral sclerosis—an immunocytochemical and ultrastructural study. Acta Neuro-pathol. 83, 579-583.

Mayall, F. G., Goddard, H. & Gibbs, A. R. (1992). Intermediate filament expression in mesotheliomas—leiomyoid mesotheliomas are not uncommon. Histopathology 21, 453- 457.

McLean, W.H.I. & Lane, E.B. (1995). Intermediate filaments in disease. Curr. Op. Cell Biol. 7, 118-125.

McConnell, S. J. & Yaffe, M. P. (1992). Nuclear and mitochondrial inheritance in yeast depends on novel cytoplasmic structures defined by the MDMI protein. J. Cell Biol. 118, 385-395.

McConnell, S. J. & Yaffe, M. P. (1993). Intermediate filament formation by a yeast protein essential for organelle inheritance. Science 260, 687-689.

McMahon, R. F. T., McWilliam, L. J., & Mosley, S. (1992). Evaluation of three techniques for differential diagnosis of prostatic needle biopsy specimens. J. Clin. Pathol. 45, 1094-1098.

McTavish, C. F., Nelson, W. J., & Traub, P. (1983). The turnover of vimentin in Ehrlich acites tumor cells. FEBS Lett. 154, 251-256.

Miettinen, M. (1992). Neuroendocrine differentiation in adrenocortical carcinoma—new immunohistochemical findings supported by electron microscopy. Lab. Invest. 66, 169-174.

Migheli, A., Pezzulo, T., Attanasio, A. & Schiffer, D. (1993). Peripherin immunoreactive structures in amyotrophic lateral sclerosis. Lab. Invest. 68, 185-191.

Mizutani, M., Nunoya, T., Yamasaki, H. & Itakura, C. (1992). The hypotrophic axonopathy mutant in Japanese quail. J. Hered. 83, 234-235.

Molenaar, W. M., Oosterhuis, J. W., Oosterhuis, A. M. & Ramaekers, F. C. S. (1985). Mesenchymal and muscle-specific intermediate filaments (vimentin and desmin) in relation to differentiation in childhood rhabdomyosarcomas. Hum. Pathol. 16, 838-843.

Moll, R., Franke, W. W., Schiller, D. L., Geiger, B. & Krepler, R. (1982). The catalog of human cytokeratins: patterns of expression in normal epithelia, tumors and cultured cells. Cell 31, 11-24.

Moll, R., Krepler, R., & Franke, W. W. (1983). Complex cytokeratin polypeptide patterns observed in certain human carcinomas. Differentiation 23, 256-269.

Moll, R., Lowe, A:, Laufer, J., & Franke, W. W. (1992). Cytokeratin-20 in human carcinomas—a new histodiagnostic marker detected by monoclonal antibodies. Am. J. Pathol. 140, 427-447.

Muma, N. A., Slunt, H. H., & Hoffman, P. N. (1991). Postnatal increases in neurofilament gene expression correlate with the radial growth of axons. J. Neurocytol. 20, 844-854.

Murti, K. G., Goorha, R., & Klymkowsky, M. W. (1988). A functional role for intermediate filaments in the formation of frog virus 3 assembly sites. Virology 162, 264-269.

Murti, K. G., Kaur, K., & Goorha, R. M. (1992). Protein kinase-C associates with intermediate filaments and stress fibers. Exp. Cell Res. 2O2, 36-44.

Nazeer, T., Barada, J. H., Fisher, H. A. G., & Ross, J. S. (1991). Prostatic carcinosarcoma—case report and review of literature. J. Urol. 146, 1370-1373.

Neumann, P. E., Horoupian, D. S., Goldman, J. E., & Hess, M. A. (1984). Cytoplasmic filaments of Crooke's hyaline change belong to the cytokeratin class. Am. J. Path 116, 214-222.

Neznanov, N. S. & Oshima, R. G. (1993). cis regulation of the keratin 18 gene in transgenic mice. Mol. Cell Biol. 13, 1815-1823.

Ngai, J., Coleman, T. R., & Lazarides, E. (1990). Localization of newly synthesized vimentin subunits reveals a novel mechanism of intermediate filament assembly. Cell 60, 415- 427.

Nishizawa, K., Yano, T., Shibata, M., Ando, S., Saga, S., Takahashi, T., & Inagaki, M. (1991). Specific localization of phosphointermediate filament protein in the constricted area of dividing cells. J. Biol Chem. 266, 3074-3079.

Norton, A. J., Thomas, J. A., & Isaacson, P. G. (1987). Cytokeratin-specific monoclonal antibodies are reactive with tumors of smooth muscle derivation. An immunocytochemical and biochemical study using antibodies to intermediate filament cytoskeletal proteins. Histopathology 11, 487-499.

O'Connor, C. M., Gard, D. L., & Lazarides, E. (1981). Phosphorylation of intermediate filament proteins by cAMP-dependent protein kinases. Cell 23, 135-143.

Ogawa, H., Sato, Y., Takeshita, I., Tateishi, J., & Kitamura, K. (1985). Transient expression of glial fibrillary acidic protein in developing oligodendrocytes in vitro. Brain Res. 350, 133-141.

Ohara, O., Gahara, Y., Miyake, T., Teraoka, H., & Kitamura, T. (1993). Neurofilamcnt deficiency in Quail caused by nonsense mutation in neurofilament-L gene. J. Cell Biol. 121, 387-395.

Omary, M. B., Baxter, G. T., Chou, C. F., Riopel, C. L., Lin, W. Y., & Strulovici, B. (1992). PKC ε- related kinase associates with and phosphorylates cytokeratin 8 and 18. J. Cell Biol. 117, 583-593.

Osborn, M., Altmannsberger, M., Debus, E., & Weber, K. (1985). Differentiation of the major human tumor groups using conventional and monoclonal antibodies specific for individual intermediate filament proteins. Ann. N. Y. Acad. Sci. 455, 649-668.

Osborn, M., Hill, C., Altmannsberger, M., & Weber, K. (1986). Monoclonal antibodies to titin in conjunction with antibodies to desmin separate rhabdomyosarcomas from other tumor types. Lab. Invest. 55, 101-108.

Osborn, M. & Weber, K. (1983). Biology of disease. Tumor diagnosis by intermediate filament typing: a novel tool for surgical pathology. Lab. Invest. 48, 372-394.

Osinska, H.E. & Lemanski, L.F. (1989). Immunofluorescent localization of desmin and vimentin in developing cardiac muscle of Syrian hamster. Anat Rec. 223. 406-413.

Oudega, M. & Marani, E. (1991). Expression of vimentin and glial fibrillary acidic protein in the developing rat spinal cord: an immunocytochemical study of the spinal cord glial system. J. Anat. 179, 97-114.

Parham, D. M., Webber, B., Holt, H., Williams, W. K. & Maurer, H. (1991). Immunohistochemical study of childhood rhabdomyosarcomas and related neoplasms—results of an intergroup rhabdomyosarcoma study project. Cancer 67, 3072-3080.

Parry, D. A. & Steinert, P. M. (1992). Intermediate filament structure. Curr. Opin. Cell Biol. 4, 94-98.

Paulin-Levasseur, M., Scherbarth, A., Traub, U., & Traub, P. (1988). Lack of lamins A and C in mammalian hemopoietic cell lines devoid of intermediate filament proteins. Eur. J. Cell Biol. 47, 121-131.

Pekny, M., Leveen, P., Pekna, M., Eliasson, C., Westermark, B., & Betsholtz, C. (1995). Mice deficient for glial fibrillary acidic protein (GFAP) generated by gene targeting. J. Cell. Biochem. Suppl. 19B, 139 (abstract).

Pette, D. & Staron, R. S. (1990) Cellular and molecular diversities of mammalian skeletal muscle fibers. Rev. Physiol. Biochem. Pharmacol. 116, 1-76.

Pollanen, M. S., Bergeron, C., & Weyer, L. (1992). Detergent-insoluble cortical Lewy body fibrils share epitopes with neurofilament and tau. J. Neurochem. 58, 1953-1956.

Pollanen, M. S., Dickson, D. W., & Bergeron, C. (1993). Pathology and biology of the Lewy body. J. Neuronathol. Exp. Neurol. 52, 183-191.

Porte, A., Stoeckel, M., Sacrez, A., & Batzenschlager, A. (1980). Unusual familial cardiomyopathy with storage of intermediate filaments in the cardiac muscular cells. Virchows Arch. 386, 43-58.

Potschka, M., Nave, R., Weber, K., & Geisler, N. (1990). The two coiled coils in the isolated rod domain of the intermediate filament protein desmin are staggered. A hydrodynamic analysis of tetramers and dimers. Eur. J. Biochem. 190, 503-508.

Pruss, R. M., Mirsky, R., & Raff, M. C. (1981). All classes of intermediate filaments share a common antigenic determinant defined by a monoclonal antibody. Cell 27, 419-428.

Raats, J. M., Pieper, F. R., Vree, E. W., Verrijp, K. N., Ramaekers, F. C., & Bloemendal, H. (1990). Assembly of amino-terminally deleted desmin in vimentin-free cells. J Cell Biol. 111, 1971-1985.

Ramaekers, F. C. S., Haag, D., Kant, A., Moesker, O., Jap, P. H. K., & Vooijs, G. P. (1983). Coexpression of keratin- and vimentin-type intermediate filaments in human metastatic carcinoma cells. Proc. Natl. Acad. Sci. USA 80, 2618-2622.

Rappaport, L., Contard, F., Samuel, J. L., Delcayre, C., Marotte, F., Tome, F. & Fardeau, M. (1988). Storage of phosphorylated desmin in a familial myopathy. FEBS Lett. 231, 421-425.

Ripps, H. & Witkovsky, P. (1985). Neuron-glia interactions in the brain and retina. Prog. Res. 4, 181-219.

Robert, F., Pelletier, G. & Hardy, J. (1978). Pituitary adenomas in Cushing's disease: A histologic, ultrastructural and immunocytochemical study. Arch. Pathol. Lab. Med. 102, 448-455.

Rosevear, E. R., McReynolds, M. & Goldman, R. D. (1990). Dynamic properties of intermediate filaments—disassembly and reassembly during mitosis in baby hamster kidney cells. Cell. Motil. Cytoskeleton. 17, 150-166.

Rothnagel, J. A., Dominey, A. M., Dempsey, L. D., Longley, M. A., Greenhalgh, D. A., Gagne, T. A., Huber, M., Frenk, E., Hohl, D. & Roop, D. R. (1992). Mutations in the rod domains of keratins 1 and 10 in epidermolytic hyperkeratosis. Science 257, 1128-1130.

Sakakibara, S., Sekiguchi, M., Konno, S., & Kusumoto, M. (1970). Idiopathic postpartum cardiomyopathy: report of a case with special reference to its ultrastructural changes in the myocardium as studied by endomyocardial biopsy. Am. Heart J. 80, 385-395.

Salisbury, J. R. & Isaacson, P. G. (1985). Synovial sarcoma: an immunohistochemical study. J. Pathol, 147, 49-57.

Samuel, J. L., B. Jockusch, S.B. Bertier, B. Escoubet, F. Marotte, B. Swynghedauw, & Rappaport, L. 1985. Myofibrillar organization and desmin in rat heart myocytes. Basic Res. Cardiol. 2, 19-22.

Sarria, A. J., Nordeen, S. K., & Evans, R. M. (1990). Regulated expression of vimentin CDNA in cells in the presence and absence of a preexisting vimentin filament network. J. Cell 111, 553-565.

Sarria, A. J., Panini, S. R., & Evans, R. M. (1992). A functional role for vimentin intermediate filaments in the metabolism of lipoprotein-derived cholesterol in human SW-13 cells. J. Biol. Chem. 267, 19455-19463.

Sarthy, P. V. & Fu, M. (1989). Transcriptional activation of an intermediate filament protein gene in mice with retinal dystrophy. DNA 8, 437-446.

Sax, C. M., Farrell, F. X., Tobian, J. A., & Zehner, Z. E. (1988). Multiple elements are required for expression of an intermediate filament gene. Nuc. Acids Res. 16, 8057-8076.

Sax, C. M., Farrell, F. X., Zchner, Z. E., & Piatigorsky, J. (1990). Regulation of vimentin gene expression in the ocular lens. Dev. Biol. 139, 56-64.

Schliwa, M. (1986). The Cytoskeleton: An Introductory Survey. Springer Verlag, New York.

Schroeder, T. E. & Otto, J. J. (1991). Snoods: a periodic network containing cytokeratin in the cortex of starfish oocytes. Devel. Biol. 144, 240-247.

Schultheiss, T., Lin, Z. X., Ishikawa, H., Zamir, I., Stoeckert, C. J., & Holtzer, H. (1991). Desmin/vimentin intermediate filaments are dispensable for many aspects of myogenesis. J. Cell Biol. 114, 953-966.

Schwarz, M. A., Owaribe, K., Kartenbeck, J., & Franke, W. W. (1990). Desmosomes and hemidesmosomes: constitutive molecular components. Ann. Rev. Cell Biol. 6, 461-491.

Shafiq, S. A., Sande, M. A., Carruthers, R. R., Killip, T., & Milhorat, A. T. (1974). Skeletal muscle in idiopathic cardiomyopathy. J. Neurol. Sci. 15, 303-320.

Shea, T. B., Sihag, R. K., & Nixon, R. A. (1990). Dynamics of phosphorylation and assembly of the high molecular weight neurofilament subunit in NB2a/dl neuroblastoma. J. Neurochem. 55, 1784-1792.

Sihag, R. K., Jeng, A. Y., & Nixon, R. A. (1988). Phosphorylation of neurofilament proteins by protein kinase C. FEBS Lett. 233, 181-185.

Sihag, R. K. & Nixon, R. A. (1991). Identification of Ser-55 as a major protein kinase A phosphorylation site on the 70-kDa subunit of neurofilaments. Early tumover during axonal transport. J. Biol Chem. 266, 18861-18867.

Skalli, O., Chou, Y. H., & Goldman, R. D. (1992). Cell cycle-dependent changes in the organization of an intermediate filament-associated protein: correlation with phosphorylation by p34cdc2. Proc. Natl. Acad. Sci. USA 89, 11959-11963.

Smedts, F., Ramaekers, F., Leube, R. E., Keijser, K., Link, M., & Vooijs, P. (1993). Expression of keratin-1, keratin-6, keratin-15, keratin-16, and keratin-20 in normal cervical epithelium, squamous metaplasia, cervical intraepithelial neoplasia, and cervical carcinoma. Am. J. Pathol. 142, 403-412.

Soellner, P., Quinlan, R. A. & Franke, W. W. (1985). Identification of a distinct soluble subunit of an intermediate filament protein: Tetrameric vimentin from living cells. Proc. Natl. Acad. Sci. USA 82, 7929-7933.

Sommers, C. L., Heckford, S. E., Skerker, J. M., Worland, P., Torri, J. A., Thompson, E. W., Byers, S. W., & Gelmann, E. P. (1992). Loss of epithelial markers and acquisition of vimentin expression in adriamycin-resistant and vinblastine-resistant human breast cancer cell lines. Cancer Res. 52, 5190-5197.

Spruill, W. A., Steiner, A. L., Tres, L. L., & Kierszenbaum, A. L. (1983). Follicle-stimulating hormone dependent phosphorylation of vimentin in cultures of rat sertoli cells. Proc. Natl. Acad. Sci. USA 80, 993-997.

Spudich, A., Meyer, T., & Stryer, L. (1992). Association of the b isoform of protein kinase-c with vimentin filaments. Cell Motil. Cytoskel. 22, 250-256.

Stappenbeck, T. S. & Green, K. J. (1992). The desmoplakin carboxyl terminus coaligns with and specifically disrupts intermediate filament networks when expressed in cultured cells. J. Cell Biol. 116, 1197-1209.

Steinert, P. M. (1988). The dynamic phosphorylation of the human intermediate filament keratin I chain. J. Biol. Chem. 263, 13333-13339.

Steinert, P. M., Marekov, L. N., Fraser, R. D., & Parry, D. A. (1993). Keratin intermediate filament structure. Crosslinking studies yield quantitative information on molecular dimensions and mechanism of assembly. J. Mol. Biol. 230, 436-452.

Steinert, P. M. & Parry, D. A. D. (1985). Intermediate filaments: conformity and diversity of expression and structure. Ann. Rev. Cell Biol. 1, 41-65.

Steinert, P. M. & Roop, D. R. (1988). Molecular and cellular biology of intermediate filaments. *Ann. Rev. Biochem.* 57, 593-625.

Steinert, P. M., Wantz, M. L., & Idler, W. W. (1982). O-phosphoserine content of intermediate filament subunits. Biochemistry 21, 177-183.

Sybert, V. P., Dale, B. A., & Holbrook, K. A. (1985). Ichthyosis vulgaris: identification of a defect in synthesis of filaggrin correlated with an absence of keratohyalin granules. J. Invest. Dermatol. 84, 191-194.

Taylor-Papadimitriou, J., Wetzels, R., & Ramaekers, F. (1992). Intermediate filament protein expression in normal and malignant human mammary epithelial cells. Cancer Treat Res. 61, 355-378.

Thompson, E. W., Lippman, M. E., & Dickson, R. B. (1991). Regulation of basement membrane invasiveness in human breast cancer model systems. Mol. Cell Endocrinol. 82, C203- C208.

Thompson, E. W., Paik, S. M., Brunner, N., Sommers, C. L., Zugmaie, G., Clarke, R., Shima, T. B., Torri, J., Donahue, S., Lippman, M. E., Martin, G. R., & Dickson, R. B. (1992). Association of increased basement membrane invasiveness with absence of estrogen receptor and expression of vimentin in human breast cancer cell lines. J. Cell Physiol. 150, 534-544.

Thornell, L.E., B. Johansson, A. Eriksson, V.P. Lehto, & Virtanen, I. (1984). Intermediate filament and associated proteins in the human heart: an immunofluorescence study of normal and pathological hearts. Eur. Heart J. 5, 213-241.

Tidball, J.G. (1992). Desmin at myotendinous junctions. Exp Cell Res. 199, 206-212.

Tint, I. S., Hollenbeck, P. J., Verkhovsky, A. B., Surgucheva, I. G., & Bershadsky, A. D. (1991). Evidence that intermediate filament reorganization is induced by ATP-dependent contraction of the actomyosin cortex in permeabilized fibroblasts. J. Cell Sci. 98, 375-384.

Tokuyasu, K. T., Maher, P. A., & Singer, S. J. (1984). Distributions of vimentin and desmin in developing chick myotubes in vivo. I. Immunofluorescence study. J. Cell Biol. 98, 1961- 72.

Tokuyasu, K. T., Maher, P. A., & Singer, S. J. (1985). Distributions of vimentin and desmin in developing chick myotubes *in vivo.* II. Immunoelectron microscopic study. J. Cell Biol. 100, 1157-1166.

Tolle, H. G., Weber, K., & Osborn, M. (1987). Keratin filament disruption in interphase and mitotic cells—how is it induced? Eur J. Cell Biol. 43, 35-47.

Torpey, N., Wylie, C. C. & Heasman, J. (1992). Function of maternal cytokeratin in Xenopus development. Nature 357, 413-415.

Troost, D., Smitt, P. A. E. S., Dejong, J. M. B. V., & Swaab, D. F. (1992). Neurofilament and glial alterations in the cerebral cortex in amyotrophic lateral sclerosis. Acta Neuropathol. 84, 664-673.

Troyanovsky, S. M., Eshkind, L. G., Troyanovsky, R. B., Leube, R. E., & Franke, W. W. (1993). Contributions of cytoplasmic domains of desmosomal cadherins to desmosome assembly and intermediate filament anchorage. Cell 72, 561-574.

Tsuda, T., Griendling, K. K., & Alexander, R. W. (1988). Angiotensin II stimulates vimentin phosphorylation via a Ca^{2+}-dependent, protein kinase C-independent mechanism in cultured vascular smooth muscle cells. J. Biol. Chem. 263, 19758-19763.

Turner, B. M. & Ruane, M. (1985). Immunofluorescent localization of cytokeratin antigens in mitotic HeLa cells using monoclonal antibodies. Eur. J. Cell Biol. 36, 48-57.

Vassar, R., Coulombe, P. A., Degenstein, L., Albers, K. & Fuchs, E. (1991). Mutant keratin expression in transgenic mice causes marked abnormalities resembling a human genetic skin disease. Cell 64, 365-380.

Venetianer, A., Schiller, D. L., Magin, T., & Franke, W. W. (1983). Cessation of cytokeratin expression in a rat hepatoma cell line lacking differentiated functions. Nature 305, 730-733.

Way, J., Heilmich, M. R., Jaffe, H., Szaro, B., Pant, H., Gainer, H., & Battey, J. (1992). A high molecular weight squid neurofilament protein contains a lamin-like rod domain and a tail domain with lysine-serine-proline repeats. Proc. Natl. Acad. Sci. USA 89, 6963-6967.

Weber, K., Plessman, U., & Ulrich, W. (1989). Cytoplasmic intermediate filament proteins of invertebrates are closer to nuclear lamins than are vertebrate intermediate filament proteins: Sequence characterization of two muscle proteins of a nematode. EMBO J. 8, 3221-3227.

Weber, K., Riemer, D., & Dodemont, H. (1991). Aspects of the evolution of the lamin/intermediate filament protein family—a current analysis of invertebrate intermediate filament proteins. Biochem. Soc. Trans. 19, 1021-1023.

Weidenthaler, B., Hausser, I. & Antonlamprecht, I. (1993). Is filaggrin really a filament-aggregating protein in vivo. Arch. Dermatol. Res 285, 111-120.

Wernert, N., Steitz, G., & Achtstatte, T. (1987). Immunohistochemical investigation of different cytokeratins and vimentin in the prostate from the fetal period up to adulthood and in prostate carcinoma. Path. Res. Pract. 182, 617-626.

Westwood, J. T., Church, R. B., & Wagenaar, E. B. (1985). Changes in protein phosphorylation during the cell cycle of Chinese hamster ovary cells. J. Biol, Chem, 260, 10308-10313.

Wible, B. A., Smith, K. E. & Angelides, K. J. (1989). Resolution and purification of a neurofilament-specific kinase. Proc. Natl. Acad. Sci. USA 86, 720-724.

Worman, H. J. (1990). Cellular intermediate filament networks and their derangement in alcoholic hepatitis. Alcohol Clin. Exp. Res. 14, 789-804.

Xu, Z., Cork, L. C., Griffin, J. W., & Cleveland, D. W. (1993). Increased expression of neurofilament subunit NF-L produces morphological alterations that resemble the pathology of human motor neuron disease. Cell 73, 23-33.

Yamamoto, Y., Izumi, K., & Otsuka, H. (1992). An immunohistochemical study of epithelial membrane antigen, cytokeratin, and vimentin in papillary thyroid carcinoma-recognition of lethal and favorable prognostic types. Cancer 70, 2326-2333.

Yamasaki, H., Itakura, C., & Mizutani, M. (1991). Hereditary hypotrophic axonopathy with neurofilament deficiency in a mutant strain of the Japanese quail. Acta Neuropathol. (Berl.) 82, 427-434.

Yano, T., Tokui, T., Nishi, Y., Nishizawa, K., Shibata, M., Kikuchi, K., Tsuiki, S., Yamauchi, T., & Inagaki, M. (1991). Phosphorylation of keratin intermediate filaments by protein kinase C, by calmodulin-dependent protein kinase and by cAMP-dependent protein kinase. Eur. J. Biochem. 197(2), 281-290.

Zackroff, R. V. & Goldman, R. D. (1979). *In vitro* assembly of intermediate filaments from baby hamster kidney (BHK-21) cells. Proc. Natl. Acad. Sci. USA 76, 6226-6230.

Zehner, Z. E. (1991). Regulation of intermediate filament gene expression. Curr. Opin. Cell. Biol. 3, 67-74.

Zehner, Z. E., Li, Y., Roe, B. A., Patterson, B. M., & Sax, C. M. (1992). The chicken vimentin gene. Nucleotide sequence, regulatory elements, and comparison to the hamster gene. J. Biol. Chem. 262, 8112-8120.

Zehner, Z. E. & Paterson, B. M. (1983a). Characterization of the chicken vimentin gene: Single copy gene producing multiple mRNAs. Proc. Natl. Acad. Sci. USA 80, 911-915.

Zehner, Z. E. & Paterson, B. M. (1983b). Vimentin gene expression during myogenesis: two functional transcripts from a single copy gene. Nucleic Acids Res. 11, 8317-8332.

Zieve, G. W. (1985). Mitosis-specific phosphorylation of vimentin. Ann. N. Y. Acad. Sci. 455, 720-723.

Zieve, G. W., Heidemann, S. R., & McIntosh, J. R. (1980). Isolation and partial characterization of a cage of filaments that surround the mammalian mitotic spindle. J. Cell Biol. 87, 160-169.

Chapter 8

The Endoplasmic Reticulum

GORDON L.E. KOCH

Principles of Medical Biology, Volume 2
Cellular Organelles, pages 189-214

ISBN:1-55938-803-X

INTRODUCTION

The endoplasmic reticulum (ER) represents one of the major intracellular organelle systems in eukaryotic cells (Alberts et al., 1983). Its evolution conferred upon the cell a set of capacities which permitted a degree of control over a variety of essential cellular functions which was not available to the primitive cell. These include such diverse processes as the synthesis and assembly of the elements which make up the entire membrane system of a cell. These proteins generally need to be more stable than their intracellular counterparts since the extracellular milieu is more harsh and variable. Thus the proteins probably need to be assembled in the more controlled fashion to confer these special properties. The ER also permits the cell a degree of control over intracellular calcium, the isolation and disposal of toxins, as well as the contribution to the general maintenance of the integrity of cytoplasmic structure. Thus it is difficult to imagine that a satisfactory description of cellular function can be achieved without an understanding of the nature of the ER and the ways in which it performs its functions.

In this chapter the emphasis will be upon the examination of the current status of our understanding of the structural elements which make up this organelle. Considerable progress has been made in this area during recent years and a solid foundation is being laid for the analysis of ER function in structural terms.

THE NATURE OF THE ER

The term endoplasmic reticulum refers to the extensive system of intracellular membranes first seen by Porter (Porter et al., 1945; Porter, 1953; Porter, 1961) in early electron micrographs of cells. It is now clear, however, that the simple morphological criterion used to define the ER is no longer sufficient because other organelles such as the trans-Golgi network (Griffiths and Simons, 1986), lysosomes (Swanson et al., 1984), and endocytic vesicles (Hopkins et al., 1990) can assume similar configurations. Confusion between these and the smooth ER is quite possible particularly in isolated preparations since cell lysis often leads to vesiculation of the intracellular organelles and their separation becomes correspondingly difficult. The only ER elements which can be formally identified on morphological grounds are the rough elements since no other organelle is known to express the characteristic membrane profile exhibited by these. However, the absence of ribosomes is itself not a formal criterion for the elements being smooth ER or some other organelle since there is evidence that the density of ribosomes can vary significantly on the rough ER. Thus there could be areas in which the density is so low that they may appear devoid of ribosomes, although all other aspects of their structure and function may be identical to that of typical rough ER. This illustrates the fact that the ER is still a poorly defined organelle in structural terms and much needs to be done to describe it adequately.

STRUCTURAL ELEMENTS OF THE ER

Biological structures are characterized by an extraordinary degree of complexity and the ultimate test of the adequacy of a structural description of such complex system is the extent to which the spatial relationships of all the components in the system can be determined. However, studies on the structures of even relatively simple systems, such as individual proteins, show that such a goal cannot be achieved in a simple linear progression because no single technique exists which can provide all the required information. Rather, it is necessary to identify different levels of structural complexity within a system and to analyze each level separately with the aim that all the separate approaches will ultimately yield a consistent picture (Perutz, 1962). In the case of protein structure, the hierarchies are referred to as primary (composition and sequence), secondary (organization into local motifs such as helices), tertiary (independent folded globular domains within a single polypeptide chain), and quarternary (organizations of tertiary domains into oligomeric structures).

When analyzing a system such as the ER it is also useful to define a set of hierarchical structural elements similar to those described above if only because it provides a convenient framework for analysis. To this end, it is first important to define the basic structural elements in this context and I have suggested the following classification. The *primary* structural elements of the ER are the molecular components such as the proteins, lipids, and ions. The *secondary* elements are the local organizational motifs formed by the individual components. In the case of the ER it cannot be assumed that such a level does exist since it is possible that no local interactions occur and all of the primary structural elements behave as though they were independent, freely diffusing entities. However, if this were true it would probably represent the only such cellular compartment, and that seems unlikely. The *tertiary* structure of the ER represents the organization of the organelle into domains. The existence of at least two such domains in the ER, i.e., the rough and smooth, is already well established and itself provides *prima facie* evidence for the existence of a secondary structure since in the absence of such interactions the composition of the ER would be expected to equilibrate into a single homogeneous system. The *quarternary* structure of the ER refers to the existence of various morphological states of the organelle such as the cisternae, tubules, and vesicles.

We will now examine the ER within the framework described above and then determine the extent to which the various functions of the ER can be explained within the known structural features of the organelle.

PRIMARY STRUCTURE OF THE ER MEMBRANE SYSTEM

Lipids

Since the ER is a membrane system, it follows that one of the major classes of the molecular components must be the lipids which make up the typical bilayer of membranes. Isolated ER membranes stripped of peripheral components contain about 45% of the total dry weight as lipid compared with about 50% protein (Esko and Raetz, 1983). Since these are analyses of stripped membranes, the protein probably represents integral membrane proteins of the ER. However, it is not clear whether this is a representative ratio for all ER membranes or whether the lipid-to-protein ratio is different for different parts of the reticulum, e.g. the rough and smooth. The lipid-to-protein ratio of the ER membranes is similar to that of the plasma membrane but only half of that of lysosomes. However, this may reflect an anomaly in lysosomal membranes rather than a major difference between ER and plasma membranes.

One of the striking differences between the ER lipids and those of other membranes lies in the relatively low content of cholesterol and sphingomyelin in the former (Esko and Raetz, 1983). The low content of these particular lipids can be clearly traced to the rough ER since membranes from this domain can be isolated in a relatively pure form and analyzed rigorously. The Golgi apparatus contains less phosphatidylcholine and more sphingomyelin than the ER and these differences become more pronounced in the plasma membrane. A gradual increase in the amount of phosphatidylserine and decrease in the amount of phosphatidylinositol are observed when the compositions of the ER,

Golgi, and plasma membranes are compared; the ratio of cholesterol to phospholipid increases in this series.

The existence of a characteristic lipid composition in the ER is supported by studies using fluorescent probes such as the carbocyanine dyes. These agents are not completely specific but they do bind preferentially to some membranes. The dye DiOC6 labels the plasma membrane, mitochondria, and the ER membrane but apparently not others such as the lysosomes (Terasaki et al., 1984). This labeling reflects the intrinsic composition of these membranes rather than some metabolic process associated with them since the labeling is observed in fully fixed samples. Another type of probe which has been used to label cell membranes with some specificity is the fluorescent analogue of the naturally occurring lipids. ER labeling was observed with NBD-phosphatidic acid and NBD-ceramide at low temperature and at high temperature the ceramide was preferentially transferred to the Golgi apparatus. NBD-phosphorylcholine and NBD-phosphorylethanolamine showed no labeling of the ER (Pagano and Sleight, 1985). However, as in the case of DiOC6, the probes are not that specific and the exact structural implications of these results are not clear.

The origin of the differences between the lipid compositions of the ER and other membranes is not yet known. Since the ER is a site of lipid synthesis it is conceivable that the particular combination of lipids which prevails in these membranes is simply the consequence of this feature. Thus the lipids which appear to be most predominant in the ER membrane might be those which are synthesized most rapidly. A second source of the lipid profile of the ER could be the preferential export of particular lipids from the ER membrane during such processes as secretion. Thus some lipids might become concentrated at the sites of protein exit from ER and thereby lead to their depletion from the ER membrane itself. Since cholesterol is synthesized in the ER, but does not accumulate in the ER membrane, it presumably represents one of the lipids which is preferentially exported from the ER. However, it should be noted that in absolute terms the amount of cholesterol in the ER is considerably greater than that in other organelles such as the Golgi apparatus, and the relative increase in the cholesterol-to-phospholipid ratio in the latter could be achieved by selective accumulation of cholesterol in the Golgi rather than by selective export from the ER.

The main structural role of the ER lipids is to form the continuous lipid bilayer which characterizes the ER membrane. Although the general appearance of the ER membrane is similar to other membranes it does exhibit one major difference, i.e., it is considerably thinner (5 nm against 8-10 nm) than the plasma membrane (Bishop and Bell, 1985). Since 5 nm is the thickness of a lipid bilayer it appears that the ER membrane lacks the type of associated protein which confers the extra thickness upon the plasma membrane.

The structural role of the lipid in the ER membrane does not appear to extend beyond that of maintaining the integrity of the lipid bilayer. There is, however, one additional set of observations which suggests that the ER membrane might have a special arrangement of lipids. There is evidence that the ER membrane is relatively easily permeabilized to proteins in isolated microsomes (Koch et al., 1988b). Thus the extent to which the lipid in the ER membrane serves as a continuous permeability barrier may be different from that in membranes such as the plasma membrane.

Proteins

There are several categories of proteins associated with the ER. The most familiar are probably the newly synthesized proteins which are destined for transit to other parts of the secretory pathways. Because these proteins are not true ER proteins, their contribution to ER structure and function is probably insignificant. Yet, until recently, it was widely assumed that they represented the major components of this organelle (Alberts et al., 1983). This picture has changed dramatically during the last few years and it is now established that the ER contains many major resident proteins which make a significant contribution to the structure and integrity of this organelle.

Membrane Proteins

Relatively few major proteins have been identified as resident integral proteins of the ER. Much of the emphasis in this area has been directed towards the identification of proteins which might participate in the binding of ribosomes to the rough ER (RER) membrane. Thus there has been a tendency to identify RER proteins on the basis of some preferential association with ribosomes and consequently the list of prospective RER proteins identified on this basis is likely to be limited. Early studies of membrane proteins from microsomes revealed integral proteins with molecular weights of 166, 107, 100, 68, 60, 50 and 36 kD, respectively. The 60-68 kD components are probably the same as the proteins called ribophorins (Kreibich et al., 1976). The basic structure of the ribophorins appears to involve a large luminal domain suggesting that they are involved in some intraluminal function (Crimando et al., 1987). These proteins were identified as prospective ribosome binding protein, i.e., ribophorins, but subsequent studies have shown that this is unlikely (Hortsch et al., 1986). However, there are relatively large amounts of these proteins in RER preparations so they could play a significant role in the structure and function of the ER membrane.

Recent studies have identified a 180-kD protein which is probably the major ribosome binding site in the RER membrane (Savitz and Meyer, 1990). In contrast to the ribophorins, the putative ribosome receptor has a large cytoplasmic domain which contains a measurable ribosome binding capacity. This protein is the only known ER membrane protein which would be expected to be present in all RER membranes and could play an important role in ER structure, in addition to its ribosome binding function. However, the generality of the protein in different systems needs to be established.

One of the best characterized ER membrane proteins is cytochrome P-450 which has attracted much interest because of its role in drug detoxification (Sato and Omura, 1978). However, the protein has only a limited distribution in cells and therefore cannot perform a general structural role in the ER. It is possible that some ER membrane proteins perform both enzymatic and structural roles. A good example of this is HMG CoA reductase which is involved in cholesterol biosynthesis (Brown et al., 1974). It is a smooth ER protein which is highly overexpressed in cells treated with compactin (Chin et al., 1982). The overexpression of the protein is associated with a massive hypertrophy of the smooth ER and reorganization of the ER into a novel crystalloid structure (Anderson et al., 1983). The formation of this structure appears to be dependent on the overexpression of the reductase suggesting that the protein itself contributes to its formation. This is an interesting case since it suggests that gross ER morphology can be determined by specific proteins in the organelle.

From the structural point of view, a crucial question concerns the physical state of the membrane proteins in the ER. As in the case of the plasma membrane, it is important to determine whether the ER membrane proteins are freely diffusing species or whether they exist in a diffusion restricted form such as supramolecular assemblies. If the latter is true, they could make significant contributions to ER structure. In such cases it will also be important to determine whether the restricted diffusion results from interactions between the proteins within the ER membrane or whether cytoplasmic or luminal proteins contribute.

Reticuloplasmins

The term reticuloplasm is applied to the material which occupies the luminal space of the ER (Koch, 1987). It probably represents an area in ER structure and function in which most progress has been made during the last few years. The presence of a concentrated proteinaceous mileu within the ER lumen was suggested by the staining properties of the material (Kristic, 1979). Until a few years ago it was assumed that the reticuloplasm was mainly composed of newly synthesized or assembled proteins destined for other sites, particularly the extracellular milieu, as secretory proteins. The main evidence for this seemed to be the fact that the lumen was often enlarged in secretory cells or cells in which secretory proteins were accumulating. However, this was never demonstrated formally and alternative explanations, such as the presence and overexpression of a family of resident proteins which were characteristic and invariant, was not considered. Early studies on microsomal preparations (Kaderbhai and Freedman, 1980; Kaderbhai and Austen, 1984) had shown that there were proteins which were characteristic of these preparations but their general cellular location or their microlocalization were not investigated. The first clear demonstration of a set of proteins characteristically associated with the ER was provided by the studies of Green et al., who identified several proteins which appeared to be ER associated on the basis of cell fractionation analysis (Lewis et al., 1985a). However, it was concluded that the proteins were membrane proteins and one ERP99 (endoplasmin) was even identified as a transmembrane protein (Lewis et al., 1985b).

The formal demonstration of the presence of an abundant luminal ER specific protein was provided by Koch et al. with studies on the glycoprotein endoplasmin (Koch et al., 1985; Koch et al., 1988). This was shown to be an abundant glycoprotein in a wide variety of cells and species. It had the carbohydrate modification characteristic of an ER glycoprotein and was clearly localized by immunoelectron microscopy to the ER. Careful analyses showed that it was a luminal protein not an integral membrane protein (Koch et al., 1988). Since it was the first protein of this type to be formally identified it was called endoplasmin.

It was also proposed that the generic name of reticuloplasmins (Koch, 1987) should be accorded to all proteins which were characterized by being luminal and ER specific, i.e., they are the proteins which represent the true protein components of the reticuloplasm. Thus, they would be distinguished from the itinerant proteins found in the ER lumen and those which might only exist in the reticuloplasm of particular cells. The corollary of this is that the reticuloplasmins represent proteins which carry out functions which are characteristic of the ER lumen. They therefore represent a major focus for the study of ER structure and function.

The Major Reticuloplasmins

Studies on mammalian cells have indicated that four of the reticuloplasmins represent the major proteins of this compartment (Macer and Koch, 1988). They are endoplasmin, BiP, protein disulphide isomerase (PDI), and calreticulin. The clearest indication that these are the major reticuloplasmins comes from systematic studies of the ER of a plasmacytoma line (Macer and Koch, 1988). These cells contain large amounts of RER and are probably as good a source of typical ER as any. Estimates of the collective concentration of these proteins in the ER lumen indicate that they achieve levels from 30-100 mg per ml which is comparable to that of the proteins in the cytoplasm. Thus the reticuloplasm is a protein-rich medium and this factor must be taken into consideration when considering the structural and functional roles of the individual proteins.

Endoplasmin

Endoplasmin (GRP94, HSP108, ERP99) is a dimeric glycoprotein with identical protomers of ~ 95 kD. The protein from mouse (Mazzarella and Green, 1987) and chicken (Kuloma et al., 1986) has been sequenced completely, and partial sequences have been obtained from rat and hamster. There are six potential glycosylation sites in the protein but only one of these seems to be consistently glycosylated. The glycosylation appears to be typical for a protein which is permanently resident in the ER (Lewis et al., 1985b; Koch et al., 1986). It is homologous to the cytoplasmic protein HSP95, and EM studies have shown that they have a similar shape, i.e., they are relatively asymmetric elongated molecules with an axial ratio similar to proteins such as alpha actinin (Koyasu et al., 1986). It is an acidic protein and contains many stretches of acidic residues in its sequence. In common with several other reticuloplasmins, the acidic sequences tend to be concentrated at the C terminal end of the molecule. A significant consequence of the presence of these acidic residues is the capacity of the protein to bind calcium ions in the millimolar range (Koch et al., 1986; Macer and Koch, 1988).

A feature of endoplasmin is its high degree of sequence conservation between different species. This was originally suggested by the conservation of its migration position on two-dimensional polyacrylamide gels (Koch et al., 1985) and confirmed by the sequences. Thus there is only a 3.7% difference between the first 750 residues of the chicken and rabbit proteins. This degree of sequence conservation is associated with proteins which carry out a structural role in cells and, when taken with the fact that endoplasmin was present at very high concentrations in the reticuloplasm, led to the proposal that one of the functions of endoplasmin was to play a structural role such as the formation of supramolecular assemblies or networks in the ER (Koch, 1987). In this context it is worth mentioning that HSP90 and endoplasmin have been found to bind actin (Koyasu et al., 1986) so that at least in cytoplasm this family of proteins might play a role in stabilization of the cytoplasmic matrix.

BiP

This protein was first encountered during studies on the synthesis and assembly of immunoglobulins where it was found to form stable associations with the newly formed heavy chain (Bole et al., 1986). It is one of the abundant proteins in reticuloplasm and has been found in eukaryotes ranging from yeast to humans (Pelham, 1989). It is a monomeric protein and also has a cytoplasmic homologue, the much studied HSP 70 protein (Munro and Pelham, 1986). BiP is a relatively acidic protein although less so than endoplasmin. Thus it does not bind as much calcium as endoplasmin (Macer and Koch, 1988) and it is debatable whether the low level of calcium binding observed is significant. There is no evidence that it is ever glycosylated. Like endoplasmin, BiP exhibits a high degree of sequence conservation which would also favor its participating in a structural role.

It has been demonstrated that BiP has a tendency to associate with a variety of partially folded proteins (Pelham, 1986). This has led to a variety of suggestions about the role of BiP in cells. One is that BiP and its cytoplasmic homologue, HSP70, act to renature proteins which are denatured and aggregated by heat shock or stress in an ATP-dependent process. The only evidence for this so far is some studies on *in vitro* aggregates of proteins (Lewis and Pelham, 1985) and the indication that HSP70 proteins have the ability to dissociate protein assemblies such as coated vesicles (Ungewickell, 1985). Another suggestion is that BiP acts to prevent nascent or partially unfolded proteins from denaturing or aggregating. The idea seems to be that nascent chains are bound to BiP during the folding process, when they are vulnerable to aggregation, and are detached when folding is complete. A variant of this argument is that BiP assists protein folding by reversing the folding of incorrectly folded proteins in an ATP-dependent process.

In the case of oligomeric proteins, BiP is thought to bind the protomers and prevent denaturation or aggregation until the correct oligomer has assembled (Hurtley and Helenius, 1989). Yet another function proposed for HSP70, which might also be relevant to BiP, is in assisting the transport of polypeptide chains across the ER membrane. One view is that the cytoplasmic proteins are unfolded by HSP70 and thereby permitted to pass through an ER membrane channel. Another is that HSP70 or a similar molecule actually drives the unfolded polypeptide through the membrane pore in an energy driven process. BiP could perform a complementary role on the luminal side of the ER.

The plethora of putative functions associated with BiP and HSP70 raises the question of whether they are multifunctional proteins or whether there is one basic property which accounts for all these functions. Pelham (1986) has proposed that the common factor is the ability of BiP and HSP70 to bind to exposed hydrophobic regions and to reverse this process with ATP. The implication of such a model is that proteins must possess a variety of BiP binding sites. Thus it is surprising that in the best studied system, Ig assembly, it has been found that BiP binding to the heavy chain is localized to a single domain. Since the separate domains are thought to fold independently, some folding appears to occur in the absence of BiP in this system. Furthermore, posttranslational folding in a microsomal system has been shown to occur efficiently in the absence of BiP (Bulleid and Freedman, 1988). A more likely explanation is that the BiP binding sites are specific regions introduced into the protein during evolution to actively assist the folding of particular proteins in a BiP (or HSP70) dependent fashion. These sites do not need to be hydrophobic but rather they would represent segments which are crucial in the folding pathway. It is worth noting here that in evolutionary terms BiP may not have originally developed for this function. Rather BiP binding sites might have evolved independently and convergently to assist proteins to exploit the special properties of this particular protein. Clearly the central issue which needs to be addressed in relation to the role of BiP is the identity of the binding sites on polypeptides, i.e., are they of some general type or are they specific sequences which have evolved for BiP binding?

Protein Disulphide Isomerase (PDI)

This protein was discovered many years ago as aiding an activity in cells which could catalyze the formation of disulphide bonds in proteins (De Lorenzo et al., 1966). This activity is associated with a homodimer of ~55 kD protomers (Freedman, 1989). However, the situation has become more complicated recently with the discovery of a number of other activities associated with the homodimer as well as the realization that there are heteromeric species in which only one of the subunits is that found in the homodimer. Thus the prolyl-4-hydroxylase beta subunit has been shown to be identical to PDI (Koivu et al., 1987).

A protein which is thought to be involved in oligosaccharide transfer to the glycosylation sites of proteins also appears to be identical to PDI (Geetha-Habib et al., 1988). Other proteins which have apparent similarities or identities with PDI are thyroid hormone binding protein (Cheng et al., 1987) and phosphoinositide-specific phospholipase C (Bennett et al., 1988). It should, however, be emphasized that the major species in well characterized systems is the familiar PDI homodimer (Macer and Koch, 1988) and the relevance of many of the other species in terms of ER structure remains to be determined.

With respect to function, the main role of PDI appears to be to catalyze the reshuffling of disulphide bonds until the correct configuration is achieved (Freedman, 1984). It has also been suggested that PDI might also perform a role analogous to that proposed for BiP, i.e., to keep proteins in an unfolded state until various modifications have been carried out (Freedman, 1989). Presumably such a mechanism would operate by delaying the formation of the fully folded conformation. However, such a suggestion seems to imply that PDI slows down protein folding whereas the traditional view is that it helps folding to occur more efficiently and speedily.

Calreticulin

This is a ~55 kD reticuloplasmin which is monomeric when isolated (Macer and Koch, 1988). Its most striking characteristic is its exceptional acidity and calcium binding which led to the suggestion that it could be involved in calcium storage within the ER (Koch, 1987; Macer and Koch, 1988). The existence of the protein in reticuloplasm was first demonstrated in a systematic study of isolated reticuloplasm (Macer and Koch, 1988) and confirmed by immunofluorescence and immunoelectron microscopy (Koch et al., 1989a). The complete amino acid sequence has been determined (Smith and Koch, 1989; Fleigel et al., 1989). Subsequently, it has emerged that calreticulin is the same protein as a high affinity calcium binding protein identified in sarcoplasmic reticulum (Ostwald and MacLennan, 1974) and a so-called calregulin (Khanna et al 1986). Both these studies emphasized the presence of high affinity calcium binding sites but the functional significance of this is not clear. However, the presence of the protein in the sarcoplasmic reticulum adds further credence to the putative calcium storage function.

The primary structure of calreticulin displays a number of interesting properties. Proteolytic fragmentation studies suggested that the protein might have a zonal character since it was possible to generate stable fragments by proteolysis (Macer and Koch, 1988). Furthermore, a large proportion of the acidic residues and the calcium binding activity were associated with the C-terminus (Koch et al., 1989a). The zonal character of the protein was confirmed by the complete sequence which showed an amino terminal neutral region followed by a central proline rich zone with repeated sequences and a polyacidic carboxy terminal zone. Broadly speaking, the sequence suggested a globular N-terminal domain followed by an extended acidic C-terminal domain (Koch et al., 1989b) A short sequence within the protein is clearly homologous with a sequence in calsequestrin. An interesting question relates to the existence of a cytoplasmic homologue of calreticulin in line with the demonstration that endoplasmin and BiP have such homologues. Some evidence for this has come from studies on an invertebrate species where a protein which is clearly homologous with calreticulin has been identified (Unnasch et al., 1988). However, the protein lacks the acidic C-terminal region and the KDEL sequence associated with the ER molecule. Thus it could represent a cytoplasmic homologue of calreticulin.

One aspect of calreticulin which has caused much confusion is its relationship with calsequestrin. Interest in this area has been driven by the wider issue of calcium storage in nonmuscle cells. Much circumstantial evidence has developed for the thesis that the ER is the major source of calcium for intracellular signaling, thus performing the role of the sarcoplasmic reticulum in nonmuscle cells (Carafoli, 1987). Support for this was first obtained with the demonstration that endoplasmin could bind significant amounts of calcium in the low affinity range (Koch et al., 1986). Further analyses of reticuloplasm showed that calreticulin was the other major calcium binding species with modest binding by BiP and PDI (Macer and Koch, 1988). In spite of the considerable superficial similarities between calreticulin and calsequestrin, even preliminary sequencing and immunochemical analyses showed they were different proteins (Koch, 1987; Macer and Koch, 1988; Koch 1989a), which was confirmed by the complete sequence.

However, contemporaneous with these developments came the claim that nonmuscle cells contained a calsequestrin like protein which was localized not in the ER but in a novel organelle, the calciosome, which was proposed as the true calcium storage organelle (Volpe et al., 1988). This was puzzling in view of the previous studies which has virtually eliminated the existence of a calsequestrin homologue in nonmuscle cells. It has now been confirmed that the protein marker used originally to define the calciosome was actually calreticulin (Koch, 1990; Treves et al., 1990). This fact alone raises doubts about the existence of calciosomes since it is established that calreticulin is a reticuloplasmin. Other evidence on the localization of other proteins involved in calcium signaling also supports the idea that these proteins are localized in the ER and not in some separate organelle (Nigan and Towers, 1990).

Other Reticuloplasmins

The four proteins described above represent the four major reticuloplasmins expressed in the ER of cells such as plasmacytomas which are specialized for secretion. The only other major candidate reticuloplasmin in such cells is a protein with a mobility slightly slower than PDI on SDS polyacrylamide gels (Macer and Koch, 1988) and hence referred to as RP60 (60 kD reticuloplasmin). The protein has not been extensively analyzed, although its N-terminal sequence is not found in any of the known proteins.

Several other proteins also appear to be located in the ER as evidenced by the presence of the KDEL sequence at the C-terminus (Pelham, 1989). However, it is not yet formally established whether the presence of a KDEL sequence is a sufficient criterion for the identification of a reticuloplasmin and direct localization will be required.

In summary, it is now apparent that the reticuloplasm is characterized by the presence of a unique family of proteins, the reticuloplasmins, which make up the principal structural elements of this compartment.

Ionic Composition of Reticuloplasm

The ionic components of the ER remain a matter of some controversy, largely because there is no simple way of measuring these directly *in situ*. It is, however, generally accepted that the most reliable method available is electron probe X-ray microanalysis. This method is particularly suitable for the analysis of calcium ions and it has been used to demonstrate that the calcium ion concentration in the rough ER approaches the millimolar level (Somlyo et al., 1985). However, this interpretation has been challenged on the grounds that the probes used are relatively large and the true calcium stores could be smaller structures, such as the putative calciosomes, which would not be resolved by the probes used (Meldolesi et al., 1988). This speculation appears to have been refuted by a more recent study (Baumann et al., 1991). From the functional standpoint no particular role has been ascribed to the other major ions such as potassium, magnesium, and phosphate in the ER.

SECONDARY STRUCTURE OF THE ER MEMBRANE SYSTEM

It is a measure of the scale of ignorance we possess about the fundamental structural characteristics of the ER that such a basic question as the existence of a secondary structure has not even been addressed until recently. The common, albeit unproven, assumption is that ER proteins, especially those in the reticuloplasm, are soluble, freely-diffusing species (Pelham, 1989). However, it was proposed sometime ago that the high concentrations of these proteins and their sequence conservation were consistent with a structural role (Koch, 1987), i.e., the proteins could be associated into a luminal matrix rather like that existing in various other compartments such as the periplasm, sarcoplasmic reticulum, and cytoplasm itself. Such an arrangement could be of considerable importance to the structural stability of the ER which displays substantial variability in this respect from one cell to another. A luminal matrix could provide additional stability and rigidity in those cases where this is important, e.g., in cells exposed to calcium dependent stress (Schanne et al., 1979). It could provide a primary mechanism by which the reticuloplasmins are selectively retained in the ER with the KDEL based system serving to complement this bulk sorting process (Booth and Koch, 1989). The calcium storage function associated with the ER would also be served by a structure since a structured matrix appears to be an essential feature of the functioning of the sarcoplasmic reticulum (Meissner, 1975). It could also provide a device by which the ER could assume different domains, such as the rough and smooth, as well as the regions which appear to be specialized for virus assembly.

Although the idea of a luminal matrix has been questioned (Pelham, 1989), both directly and by implication, there has been considerable circumstantial support for the proposed existence of a luminal matrix. Apart from the fact that most compartments with high concentrations of conserved proteins do indeed have them organized into a structure, studies on the diffusion rates of reticuloplasmins are inconsistent with the concept of free diffusion (Ceriotti and Coleman, 1988). Thus when BiP and ovalbumin were expressed in frog oocytes and their movement through the ER measured, it was found that the rate of diffusion of BiP was indeed slower than that of ovalbumin. In a similar vein, studies on the secretion of BiP depleted the DKEL retention sequence showed that it was secreted at a rate substantially slower than lysozyme (Munro and Pelham, 1987). It was assumed that this reflected some interaction with the ER membrane (Warren, 1987). An alternative is that it reflects associations between the proteins which restricts their diffusion in the ER. The observation that the stability of ER in the presence of calcium ionophores is increased when the reticuloplasmins are overexpressed during the stress response could be reconciled with a structural role for the proteins (Koch et al., 1988). The induced secretion of the reticuloplasmins by longer-term exposure to calcium ionophore and its reversal by the stress response has also been interpreted in terms of the existence of an intraluminal matrix of reticuloplasmins which can be stabilized by calcium ions (Booth and Koch, 1989).

Preliminary evidence for an association between the reticuloplasmins in intact cells comes from analyses of the fixation efficiency of the proteins with formaldehyde (G. Koch, unpublished observations). It was found that the proteins in the reticuloplasm of intact cells could be efficiently cross-linked into a supramolecular aggregate under conditions which did not even cross link soluble proteins *in vitro*. A limitation of these studies in that they do not formally exclude the possibility that the conditions in the ER are completely different from those which can be reasonably simulated *in vitro* although the observed efficiency is difficult to explain in such terms. A more direct approach to this would be to measure the actual diffusion rates of the proteins in the ER but this is not yet technically feasible. However, it should be possible to use the approach used by Schliwa et al. (1981) to examine whether the proteins are extracted when the ER membrane is rendered permeable to proteins with agents such as detergents. In view of the fundamental importance of this issue, and its relevance to the various structural and functional questions discussed above, this remains one of the central limitations to a proper understanding of the nature of the ER.

TERTIARY STRUCTURE OF THE ER MEMBRANE SYSTEM

The existence of domains within the ER was established at a very early stage in the study of this organelle. The two major domains were morphologically distinct since one was studded with ribosomes and the other was not. Systematic studies established that the two regions were connected so it was clear that they represented separate domains of a single continuous organelle. Subsequent studies also confirmed that the two domains were functionally different.

The structural question, which has remained unanswered for so long is: "What is responsible for the differences between these domains?" At the primary level it appears that there are proteins which are localized in one domain or the other. These include the ribophorins, the putative ribosomal receptor, and HMG CoA reductase. However, the mere presence of these proteins is not sufficient to define the differences between the two domains since the membranes are thought to be intrinsically fluid. Thus, there must be mechanisms which restrict the fluidity of the membrane to prevent the equilibration of these proteins. One possibility is that this is achieved by a self association mechanism involving the integral membrane proteins. However it must also be noted that the distribution of the reticuloplasmins may not be uniform and there are regions which are either enriched for or depleted of some of the proteins (Bole et al., 1989; Koch, 1990). As pointed out above, this provides a plausible mechanism for establishing the domains, i.e., the formation of a self-assembling luminal structure which could form a scaffold to which other proteins and membrane proteins could become attached. The precedent for this would be the effect of the cytoskeletal proteins on the distribution of plasma membrane proteins.

The other major structural domain associated with the ER is the tertiary or transitional zone (Jamieson and Palade, 1968). This represents a region which is a morphological hybrid between the rough and smooth since one face appears to be rough and the other, which faces the Golgi complex, is smooth. Extensions from the latter face appear to be involved in transfer of proteins from the ER to the Golgi. There is some uncertainty of the nature of these extensions since they might be vesicles, a continuous tubular reticular, or a combination of both. Whatever the details of this issue, the relevant question regarding the ER itself concerns the origin of the transitional zone. One possibility is that there are specific structural elements such as proteins which self assemble and give this region its special properties. However, a more plausible explanation is that the smooth zone is induced through a proximity to the Golgi apparatus. This would explain the observation that ER to Golgi transfer can occur in *in vitro* systems in which the ER was vesiculated (Beckers and Balch, 1989) and the observation that such transfer can also occur in an *in vivo* system in which the ER appeared to be completely vesiculated (Booth and Koch, 1989). Thus rough ER vesicles which approached the Golgi would be converted into smooth structures as a result of the inductive effect of the Golgi and fusion could ensue. Analysis of this question should be possible with an *in vitro* system which carries out efficient ER to Golgi transfer.

VIRAL ASSEMBLY DOMAINS

It has been frequently observed that the budding of virus particles into the ER occurs at particular sites rather than throughout the ER. In some cells this is associated with a morphologically distinct site and there is evidence that the composition of the reticuloplasm in such regions is different (Bole et al., 1989, Koch et al., 1989a). The question this raises is whether the site of viral assembly is a preexisting one or whether it is induced by the virus particles or proteins. It is difficult to imagine how such a specialization might be induced if the ER is a homogeneous structure since viral particles would have an equal probability of budding anywhere. It could be argued that they do bud randomly but accumulate at the observed site. In such a case it is surprising that only one structure is observed in the majority of cells in a population. It has also been claimed that a distinct site can be detected in uninfected cells (Tooze et al., 1984) but in the absence of a suitable marker this conclusion is debatable. However, there seems to be a reasonable *prima facie* case for believing that the existence of the viral assembly sites represents a natural domain within the ER of at least some cells.

QUARTERNARY STRUCTURE OF THE ER MEMBRANE SYSTEM

The term quarternary structure is used to denote the major morphological elements, i.e., the sac-like cisternae, the tubulovesicular reticulum, and the vesicular structures associated with the ER in a variety of cells. Although the elements appear to be interchangeable they do predominate in particular cell types and this presumably reflects the functional advantage of each. The cisternal elements predominate in cells such as fibroblasts. The reason for this is not clear and it may simply reflect differences in the cytoplasmic architecture of the cells. For example, the formation of the tubular reticulum has been shown to depend on microtubules in both intact cells as well as in *in vitro* models of the ER. The presence of large numbers of microtubules in some cells could favor the formation of a tubular reticulum by increasing the frequency of contacts between the ER and microtubules (Lee and Chen. 1988; Dabora and Sheetz, 1988). If this were the case the morphological elements might not be directly related to a particular ER function.

The significance of the vesicular elements of the ER is still obscure. The existence of ER associated vesicles was first demonstrated in the earliest micrographs and subsequent studies have shown that vesiculation frequently occurs in cells undergoing mitosis. *In vitro* studies on ER assembly also revealed a clear tendency for the ER to vesiculate and reform. *In vivo* dissociation was induced in some cells with calcium ionophores and the fragments were able to reassemble into continuous elements in a calcium dependent process. The tendency of the ER to dissociate into small uniform fragments was first noted by Claude (1943) even before the formal discovery of the organelle. Unfortunately, the term microsome is still used in this connection although a microsomal preparation is invariably a mixture of various types of vesicles from organelles. It was therefore suggested that the term reticulosome be used to denote the small uniform vesicles which form when the ER itself fragments (Koch et al., 1988a). The reason for the tendency of the ER to fragment into uniform vesicles is not clear since other membranes such as the plasma membrane do not exhibit this property. It is possible that such tendency might be the only way a tubular reticulum can dissociate because of the limitations of lipid availability. However, it is less obvious why the cisternal elements should do this since they cannot dissociate by pinching off. The implication is that there is some intrinsic structural characteristic general to all ER elements which causes this pattern of fragmentation. One possibility is that it reflects some periodicity in the reticuloplasm. Such an arrangement might be caused by the luminal structure previously described. In this context, it is interesting that the uniform beading which occurs in certain filopodial extensions has been related to the condensation of the actin filaments within the projection. Although this is a speculative suggestion, it does fit well with the concept of a supramolecular system in the ER lumen.

FUNCTIONS OF THE ER MEMBRANE SYSTEM

The most familiar function of the ER, which is the co-translational translocation of polypeptides across the ER membrane, has been discussed extensively over many years and will not be considered here in any detail.

Posttranslational Translocation of Polypeptides

Intense interest has developed in this subject in recent years following the demonstration that polypeptides can indeed be translocated across the ER membrane posttranslationally. This led to the realization that there must be at least two fundamental but unknown mechanisms involved. One is a mechanism to unfold the polypeptide since it needs to be threaded across the ER membrane. Evidence has been obtained that this involves the familiar HSP70 molecule which acts as an unfoldase (Deshais et al., 1988; Chirico et al., 1988). Interestingly, the claim that this was predicted (Pelham, 1986, 1988) is debatable since the original speculation was that HSP70/BiP acted to dissociate aggregates through binding to hydrophobic surfaces. In fact these proteins bind to hydrophilic peptides so their mode of action may be very different to that originally envisaged.

The other element in this process is the translocation system. The suggestion that this involves an ATP-protease type molecule seems unlikely (Rothman and Kornberg, 1986). The more convincing candidate is the family of proteins which appears to be involved in the translocation of peptides across the bacterial membrane (Townsend and Bodme, 1989). This is a particularly important system since it may have a crucial role in the formation of peptide MHC protein complexes which are required for the development of immune responses to protein antigens. Analysis of this process promises to be one of the most exciting areas in this field.

Calcium Storage by the ER

The role of the ER in calcium storage and signaling (Carafoli, 1987) has been the subject of considerable interest for many years. The central issue at the moment is whether it is the ER or an alternative organelle, the putative calciosome (Volpe et al., 1988), which carries out this function. The evidence that it is the ER is that all the prospective calcium storage proteins, notably endoplasmin and calreticulin, are localized in the reticuloplasm (Koch, 1987; Macer and Koch, 1988). Furthermore, the inositol 1,3,4-triphosphate receptor which has recently been identified has also been localized to the ER (Ross et al., 1989). The basis of the proposal for the calciosome was the presence of a calsequestrin immunologue (presumably therefore a sequence homologue) as well as the presence of a Ca^{2+}-ATPase and the IP_3 receptor in a special vesicular compartment.

It is now established that the calsequestrin immunologue is calreticulin (Koch, 1990; Treves et al., 1991), that the Ca^{2+}- ATPase may not be restricted to a special vesicular compartment (Nigam and Towers, 1990), and that the IP_3 receptor is associated with the ER. The possibility has been raised that there might be a special subcompartment of the ER which is specialized for the calcium storage/signaling function and that this represents the calciosome. The existence of such domains has been discussed but it is questionable whether it would be appropriate to refer to them as calciosomes even if they existed. Otherwise one would have to treat the rough ER as roughosome, the smooth ER as a smoothosome, and divide the sarcoplasmic reticulum into two calciosomes, one containing the storage function and the other the calcium pumping function. It is questionable whether such a concept would even be useful. Its main appeal seems to be the fact that it would satisfy the prediction that there might be more than one calcium pool in the cell. However this is itself a speculation and it seems dangerous to use one speculation to corroborate another!

The evidence that the ER does contain the prospective calcium storage proteins of nonmuscle cells prompts the issue of the extent to which the ER resembles the sarcoplasmic reticulum. First, there is no doubt that both organelles contain Ca^{2+}-ATPases. Second, the two organelles contain calreticulin. However, although calreticulin and calsequestrin can co-exist in the SR, this has not been demonstrated for the ER. Third, both organelles express receptors which could be involved in the regulation of calcium release, i.e., the IP_3 and ryanodine receptor, respectively. Furthermore, there is evidence that the reticuloplasmins are organized into a luminal structure analogous to that formed by calsequestrin within the lumen of the SR (Koch, 1990). These structural similarities, together with the numerous functional analyses, provide clear support for the hypothesis that the ER is a calcium storage and signaling organelle in nonmuscle cells. The question of other such organelles remains to be resolved.

Cytoskeletal Function

It was suggested by Porter (1953) that the ER might be suited to performing a skeletal or scaffolding function by virtue of its extensive distribution throughout the cytoplasm. However, most evidence has tended to favor the view that the morphology of the ER is determined by the cytoplasmic matrix or the microtubules or both, Studies have produced evidence that the ER itself has the required rigidity to perform a supporting role in cytoplasm (Kachar and Reese, 1988). During cytoplasmic streaming a shear force has to be transmitted between the cortical actin filaments and the internal cytoplasm. This must involve a semi-rigid continuum which permeates the cytoplasm. Evidence has been obtained. that the ER can assume a sufficiently rigid structure to transmit the force generated by the actin filaments involved. More direct evidence for a cytoskeletal role for the ER comes from studies on the ER of plasmacytoma cells (Koch et al., submitted). It has been possible to isolate cytoplasm-free models of these cells in which essentially all the cytoplasm was removed. Yet, the ER itself retained its normal morphology. Thus it is clear that at least in some cells the ER is a relatively rigid structure and this rigidity could reflect the existence of a luminal matrix of reticuloplasmins of the sort discussed above.

EPILOGUE

This chapter has attempted to examine the various aspects of the ER which make it a worthwhile subject for study. The approach has been predominantly a structural one because of the personal conviction that the only intellectually and aesthetically satisfying descriptions in biology are structural ones. In the case of the ER, a major deficiency of many of the functional studies is that they have taken place in an incomplete structural framework and an attempt has been made to rectify this by introducing some novel structural concepts Of particular importance is the idea that the ER lumen or reticuloplasm contains a gel-like matrix formed as a result of interactions between the reticuloplasmins. In one sense this is not an unexpected prediction since most other cellular compartments contain luminal matrices. What is surprising is that this was not considered until it was apparent that reticuloplasm did contain the required concentration of conserved resident luminal proteins to support such a structure. Perhaps what lies at the heart of this is the general lack of awareness of the fact that free diffusion of proteins in either the cytoplasm or membranes is rare, or to put it more graphically, "Cells Are Gels."

REFERENCES

Alberts B., Bray D., Lewis J. Raff M., Roberts K., & Watson J.D. (1983). Molecular Biology of the Cell. Garland Publishing, New York.

Anderson, R.G.W., Orci, L., Brown, M.S., Garcia Segura, L.M., & Goldstein, J.L. (1983). J. Cell Sci. 63, 1-20.

Baumann, O., Walz, B., Somlyo, A.V., & Somlyo, A.P. (1991). Proc. Natl. Acad. Sci. USA 88, 741-744.

Beckers, C.J.M., & Balch, W.E. (1989). J. Cell Biol. 108, 1245-1256.

Bell R.M., & Coleman R.A. (1980). Ann Rev. Biochem. 49, 459-487.

Bishop, W.R., & Bell, R.M. (1985). Cell, 42, 51-60.

Blobel, G. (1980). Intracellular Protein topogenesis Proc. Natl. Acad. Sci USA 77, 1496-1500.

Bole D.G., Hendershot L.J., & Kearney J.F. (1986). J. Cell Biol. 102, 1558-1566.

Bole, D.G., Down, R., Doriaux, M., & Jamieson, J.D. (1989). J. Histochem. Cytochem. 37, 1817-1823.

Booth C. & Koch G.L.E. (1989). Cell 59, 729-737.

Brown, M.S., Dana, S.E., & Goldstein, J.L. (1974). J. Biol. Chem. 249, 789-796.

Bulleid, N. & Freedman, R.B. (1988). Nature 335, 649-651.

Carafoli E. (1987). Intracellular calcium homeostasis. Ann. Rev. Biochem. 56, 395-433.

Cerriotti, A. & Coleman, A. (1988). EMBO J. 7, 633-638.

Chin, D., Luskey, K.L., Anderson, R.G.W., Faust, J.R., Goldstein, J.L., & Brown, M.S. (1982). Proc. Natl. Acad. Sci. USA 79, 1185-1189.
Brown, M.S. (1982). Proc. Natl. Acad. Sci. USA 79, 1185-1189.
Chirico, W.J., Waters, M.G., & Blobel, G. (1988). Nature 332, 805-810.
Claude, A. (1943). Science 97, 451-455.
Crimando C., Hortsch M., Gausepohl H., & Meyer D.J. (1987). EMBO J. 6, 75-82.
Dabora S.L. & Sheetz M.P. (1988). Cell 54, 37-46.
Deshais, R.J., Koch, B.D., Werner-Washbourne, M., Craig, E.A., & Shekman R. (1988). Nature 332, 801-804.
DeLorenzo, F., Goldberger, R.F. Steer, E., Givol, D., & Anfinsen C.B. (1966). J. Biol. Chem. 241, 1562-1567.
Edman J.C., Ellis L., Blacher R.W., Roth R.A., & Rutter W.J. (1985). Nature 317, 267-270.
Esko, J.D. & Raetz, C.R.H. (1983). The Enzymes. 16, 207-252.
Fleigel, I., Burns, K., MacLennan, D.H., Reithmeir, R.A.F., & Michalak, M. (1989). J. Biol. Chem. 264, 21522-21528.
Freedman, R.B. (1989). Cell 57, 1069-1072.
Freedman R.B. (1984). TIBS 9, 428-441.
Griffiths, G. & Simons K. (1986). Science 234, 438-434.
Henson J.H., Begg D.A., Beaulieu S.M., Fishkind D.J., Bouder E.M., Terasaki M., Lebeche D., & Kaminer B. (1989). J. Cell Biol. 109, 149-161.
Hortsch, M., Avossa, D. & Meyer, D.I. (1986). J. Cell Biol. 103, 241-253.
Hopkins, C.R., Gibson, A., Shipman, M., & Miller, K. (1990) Nature 346, 335-339.
Hurley, S.M. & Helenius, A. (1989). Ann Rev. Cell Biol. 5, 277-307.
Jones A.L. & Fawcett D.W. (1966). J. Histochem. Cytochem. 14, 215-232.
Kachar, B., & Reese, T.S. (1988). J. Cell Biol. 106, 1545-1552.
Kaderbhai, M.A. & Freedman, R.B. (1980). Biochem. Biophys. Acta 601, 11-21.
Khaderbhai, M.A. & Austen, B.A. (1984). Biochem. J. 217, 145-157.
Khanna, N.C. Tokuda, M., & Waisman, D.M. (1987). Biochem. J. 242, 245-251.
Khanna M.C., Tokuda, M., & Waisman, D.M. (1986). J. Biol Chem 261, 8883-8887.
Koch, G.L.E., Smith, M., & Mortara, R.A. (1985). FEBS Lett. 179, 328-331.
Koch, G., Smith, M., Macer, D., Webster, P., & Mortara, R. (1986). J. Cell. Sci 86, 217-232.
Koch, G.L.E. (1987). J. Cell Sci. 87, 491-492.
Koch, G.L.E., Booth, C., & Wooding, F.B.P. (1988a). J. Cell Sci 91, 511-522.
Koch, G.L.E., Macer, D.R.J., & Wooding, F.B.P. (1988b). J. Cell Sci. 90, 485-491.
Koch, G.L.E., Smith, M.J., Macer, D.R.J., Booth, D., & Wooding, F.B.P. (1989a). Biochem. Soc. Trans. 17, 328-331.
Koch, G.L.E., Macer, D.R.J., & Wooding, F.B.P. (1989b). Methods. Surv. Biochem. Anal. (Royal Soc. Chem.) 19, 159-165.
Koch, G.L.E. (1990) BioEssays 12, 527-531.
Koyasu, S., Nishida, E., Kaddowaki, T., Matsuzaki, F., Iida, K., Harada, F., Kasuga, M., Sakai, H., & Yahara, I. (1986) Proc. Natl. Acad. Sci. USA 83, 8054-8058.
Kreibich, G., Ulrich, B.L., & Sabatini, D.D. (1978). J. Cell Biol. 77, 464-487.
Kreibech, G., Bar-Nun, S. Cyako-Graham, M., Czichi, U., Marcantonio, E., Rosenfeld, M.G., & Sabatini, D.D. (1981). International Cell Biology 1980-1981 (Schweiger, H.G., ed.) pp. 579-589. Springer Verlag, Berlin.
Kristic, R.V. (1979). Ultrastructure of the Mammalian Cell. Springer Verlag, Berlin.

Kulomaa, M.S., Weigel, N.L., Kleinsek, D.A., Beattie, W.G., Connolly, O.M., March, C., Zaruchi-Schulz, T., Schrader, W.T., & O'Malley, B.W. (1986). Biochemistry. 25, 6244-6251.

Lee, C. & Chen, L.B. (1988). 54, 37-46.

Lewis, M.J., Mazzarella, R.A., & Green, M. (1985a). J. Biol. Chem 260, 3050-3057.

Lewis, M.J., Turco, S.J. & Green, M. (1985b). J. Biol. Chem. 260, 6926-6931.

Lewis, M.J., & Pelham, H.R.B. (1985). EMBO J. 4, 3137-3143.

Lin, J.J.C., & Queally, S.A. (1982). J. Cell Biol. 92, 108-112.

Macer, D.R.J. & Koch, G.L.E. (1988). J. Cell Sci. 92, 61-70.

Mazzarella, R.A. & Green, M. (1987). J. Biol. Chem. 262, 8875-8883.

Meldolesi, J., Volpe, P., & Pozzan, T. (1988). Trends Neurosci. 11, 449-452.

Meissner, G. (1975). Biochem. Biophys. Acta 389, 51-61.

Munro, S. & Pelham, H.R.B. (1986). Cell 46, 291-300.

Munro, S. & Pelham, H. R.B. (1987). Cell 48, 899-907.

Nigam, S., & Towers, T. (1990). J. Cell Biol. 111, 197-200.

Ostwald, T.J. & MacLennan, D.H. (1974). J. Biol. Chem. 249, 974-979.

Pagano, R.E., & Sleight, R.G. (1985). Science 229, 1051-1055.

Pathak, R.K., Luskey, K.l., & Anderson, R.G.W. (1986). J. Cell Biol. 102, 2158-2168.

Pelham, H. (1988). Nature 332, 776-777.

Pelham, H.R.B. (1986). Cell 46, 959-961.

Pelham, H.R.B. (1989). Ann Rev. Cell Biol. 5, 1-23.

Perutz, M.F. (1962). Proteins and Nucleic Acids - Structure and Function. Elsevier, Amsterdam.

Porter, K.R. (1961). J. Biophys. Biochem. Cytol. 10, 219-226.

Porter, K.R. Claude, A., & Fulham, E. (1945). J. Exp. Med, 81, 233-241.

Porter, K.R. (1953). J. Exp. Med. 97, 727-750.

Robinson, J.M., Okada, T., Castellot, J.J., & Karnovsky, M.J. (1986). J. Cell Biol. 108, 855-864.

Ross, C.A., Meldodlesi, J., Milner, T.A., Satoh, T., Supattopone, S. & Snyder, S.H. (1989). Nature 339-468-470.

Rothman, J.C. & Kornberg, R.D. (1986). Nature 322, 209-210.

Sato, R. & Ornura, T. (1978). Cytochrome. Academic Press, New York.

Savitz, A.R. & Meyer, D.I. (1990). Nature 346, 540-544.

Schanne, F.A.X., Kane, A.B., Young, E.E., & Faber, J.L. (1979). Science 206, 700-702.

Schliwa, M., Van Blerkom, J., & Porter, K.R. (1981). Proc. Natl. Acad. Sci. USA 78, 4329-4333.

Smith, M.J. & Koch, G.L.E. (1987). J. Mol. Biol. 194, 345-347.

Smith, M.J. & Koch, G.L.E. (1989). EMBO J. 8, 3581-3586.

Somlyo, A.P. (1984). Nature 309, 516-517.

Somlyo, A.P., Bond, M., & Somlyo, A.V. (1985). Nature 314, 622-625.

Sorger, P.K. & Pelham, H.R.B. (1987). J. Mol. Biol. 194, 341-344.

Struck, D. & Lennarz, W. (1980). In: The Biochemistry of Glycoproteins and Proteoglycans (Lennarz, W.J., Ed.). Plenun, New York.

Swanson, J., Bushnell, A., & Silverstein, S.C. (1984). Proc. Natl. Acad. Sci. USA 84, 1921-1925.

Terasaki, M., Song, J., Wong, J.R., Weiss, M.J., & Chen, L.B. (1984). Cell 38, 101-108.

Ting, J. & Lee, A.S. (1988). DNA 7, 275-286.

Treves, S., De Mattei, M., Lanfredi, M., Villa, A., Green, M., MacLennan, D.H., Meldolesi, J., & Pozzan, T. (1990). Biochem. J. 271, 471-480.

Trump, B.F., Beregesky, I.K., & Osornio-Vargas A.R. (1981). In: Cell Death in Biology and Pathology (Bowen I.D., & Lokshin R.A., Eds.) pp. 209-242. Chapman & Hall. London.

Tooze, J., Tooze, S.A., & Warren, G.A. (1984). Eur. J. Cell Biol. 33, 281-293.

Ungewickell, E. (1985). EMBO J. 4, 3385-3389.

Unnasch, T.R., Gallin, M.V., Soboslav, P.T., Erttmann, K.D., & Greene, B.M. (1988). J. Clin. Invest. 82, 262-269.

Volpe, P., Krause, K., Hashimoto, S., Zorzato, F., Pozzan, J., Meldolsei, J., & Lew, D.P. (1988). Proc. Natl. Acad. Sci. USA 85, 1091-1095.

Warren, G. (1987). Nature 327, 17-18.

Chapter 9

The Sarcoplasmic Reticulum

ANTHONY N. MARTONOSI

Principles of Medical Biology, Volume 2
Cellular Organelles, pages 215-251

ISBN:1-55938-803-X

INTRODUCTION

> *"Calcium has many virtues that make it quite unique among cations in its ability to complex with biological structures. Its divalency allows for a wide range of binding constants with biomolecules, its radius is compatible with peptide chelation, and its charge-to-size ratio permits it to slip into small molecular holes. Its crystal-field requirements are quite flexible, bond distances and angles are adjustable, and coordination numbers can vary from six to ten. All this gives the ion a great advantage in binding to irregular geometries of coordination sites of biological molecules that can accept the ion rapidly and sequentially and fold around it, permitting graded structural modulation. Small wonder that such an engaging character has been awarded role after role in the evolution of biological signaling!"*
>
> —Lowenstein (1989)

The cytoplasmic free Ca^{2+} concentration of living cells at rest is 10^{-7} M or less; this is 10^3 to 10^4 times lower than the free Ca^{2+} concentration of the tissue environment. The large Ca^{2+} gradient across cellular boundaries is established and maintained by powerful Ca^{2+} pumps located in the cell surface membranes (Carafoli, 1991, 1992), in the endoplasmic reticulum (Martonosi, 1983; Martonosi and Beeler, 1983; Stein, 1988, 1990; Inesi et al., 1990, 1992a,b; MacLennan, 1990; MacLennan et al., 1992; Jencks, 1992; Scarpa et al., 1992), and in the

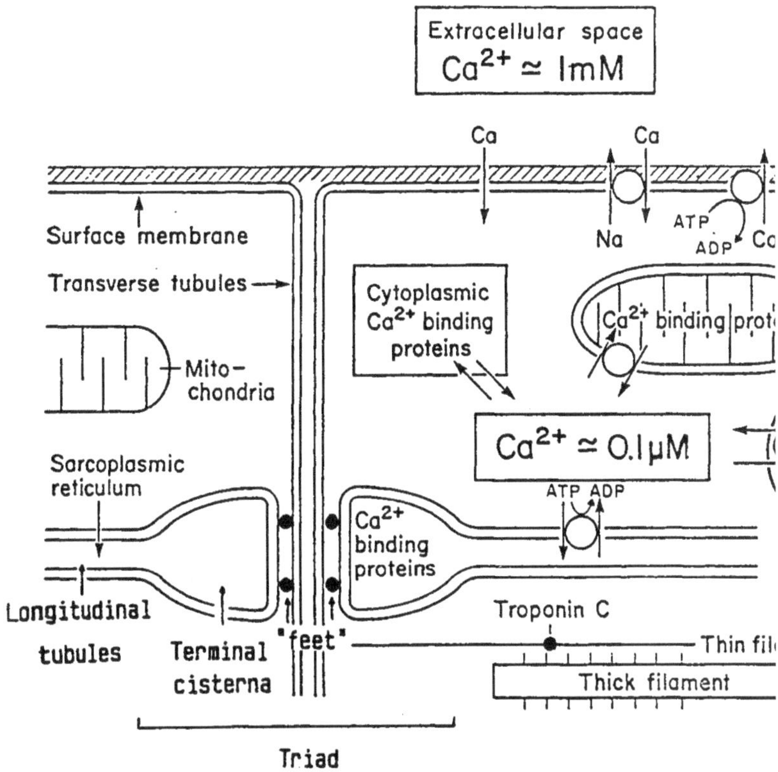

Figure 1. Scheme of systems participating in the regulation of cytoplasmic Ca^{2+} co

Table 1. Kinetic Properties of Ca^{2+} Transporting Systems in Cell Membranes

Transporting System	K_m *(Ca^{2+}) (μM)*	V_{max} *of transport (nmol Ca^{2+} per mg of membrane protein per sec, at 25°)*
Ca^{2+}-ATPase of plasma membranes	0.5	~0.5
Na^+/Ca^{2+} exchanger of plasma membranes	1-5	15-30
Ca^{2+}-ATPase of sarcoplasmic reticulum	0.5	20-30
Release route of sarcoplasmic reticulum	<1	~200-1000
Uptake route in mitochondria	1-10	~10
Na^+/Ca^{2+} exchanger of mitochondria	~10	~0.2

Notes: From Carafoli, 1988.

mitochondria (Carafoli, 1987) (Figure 1). The molecular characterictics, reaction mechanisms and Ca^{2+} affinities of the Ca^{2+} transport systems associated with these three types of cell membranes are different (Table 1; Carafoli, 1988).

In addition to the Ca^{2+} pumps, various Ca^{2+} channels, Ca^{2+} binding proteins and several ionic components of the cell also contribute to cellular Ca^{2+} homeostasis (Figure 1).

The Ca^{2+} Pump of Endoplasmic Reticulum

The endoplasmic reticulum (ER) of most eukaryotic cells contains an ATP-dependent Ca^{2+} pump of high Ca^{2+} affinity (Table 1) with a molecular weight of about 110,000 (Martonosi, 1983; Carafoli, 1987) that is distinct from the [Mg^{2+} + Ca^{2+}]-activated ATPase of surface membranes (Carafoli, 1991, 1992). The ER Ca^{2+} pump catalyzes the electrogenic transport of two Ca^{2+}/ATP hydrolyzed from the cytoplasm into the lumen of ER against a large electrochemical gradient of Ca^{2+}. Transient phosphorylation of the enzyme by ATP is an essential step in Ca^{2+} translocation. In distinction from the surface membrane Ca^{2+}-ATPase, the ER enzyme is relatively insensitive to calmodulin.

A well-characterized example of this class of Ca^{2+} transport systems is the Ca^{2+} transport ATPase of the sarcoplasmic reticulum (SR) in skeletal muscle (Martonosi and Beeler, 1983; MacLennan, 1990; Inesi et al., 1990, 1992a,b; MacLennan et al., 1992; Jencks, 1989, 1992). The principal role of the sarcoplasmic reticulum Ca^{2+} pump is the regulation of the contractile state of the muscle by accumulating much of the Ca^{2+} content of the cell in its interior during relaxation and lowering the cytoplasmic $[Ca^{2+}]$ to below 0.1 μM. During muscle contraction the Ca^{2+} is rapidly released from the sarcoplasmic reticulum, raising the cytoplasmic $[Ca^{2+}]$ to 1-30 μM with activation of contractile material (Rios and Pizarro, 1991; Rios et al., 1991, 1992).

THE STRUCTURE OF SARCOPLASMIC RETICULUM

The sarcoplasmic reticulum is a sleeve-like network of membranous tubules and cisternae that surround the myofibrils and establish junctions within each sarcomere with invaginations of the surface membranes that are called T-tubules (Figure 2; Smith, 1972). Within these junctions specialized regions of the T-tubule membranes interact on both sides with the junctional face membranes of sarcoplasmic reticulum forming the so-called triads (Figure 3; Smith, 1972). In most mammalian muscles the triads are located near the A-I junction and there are two triads per sarcomere (see Peachey and Franzini-Armstrong, 1983). This arrangement assures short diffusion distances between the contractile proteins and the sarcoplasmic reticulum.

The sarcoplasmic reticulum of skeletal muscle consists of two morphologically and functionally distinct regions: the junctional SR and the free SR (Peachey and Franzini-Armstrong, 1983).

Junctional Sarcoplasmic Reticulum

The *junctional sarcoplasmic reticulum* contains feetlike projections on its surface that interact with particles contained in the junctional region of the T-tubules, forming the *T-SR junction* (Block et al., 1988; Wagenknecht et al., 1989; Radermacher et al., 1992). These structures are involved in the transmission of the excitatory stimulus from the surface membrane to the SR, causing Ca^{2+} release (Rios and Pizarro, 1991).

Free Sarcoplasmic Reticulum

The *free sarcoplasmic reticulum* is usually divided into the *lateral sacs* and the *longitudinal tubules*, which differ in protein composition. The Ca^{2+}-ATPase is the dominant protein component, representing as much as 80% of the protein content in the SR of fast-twitch skeletal muscle; it is evenly distributed throughout the free SR. The lateral sacs

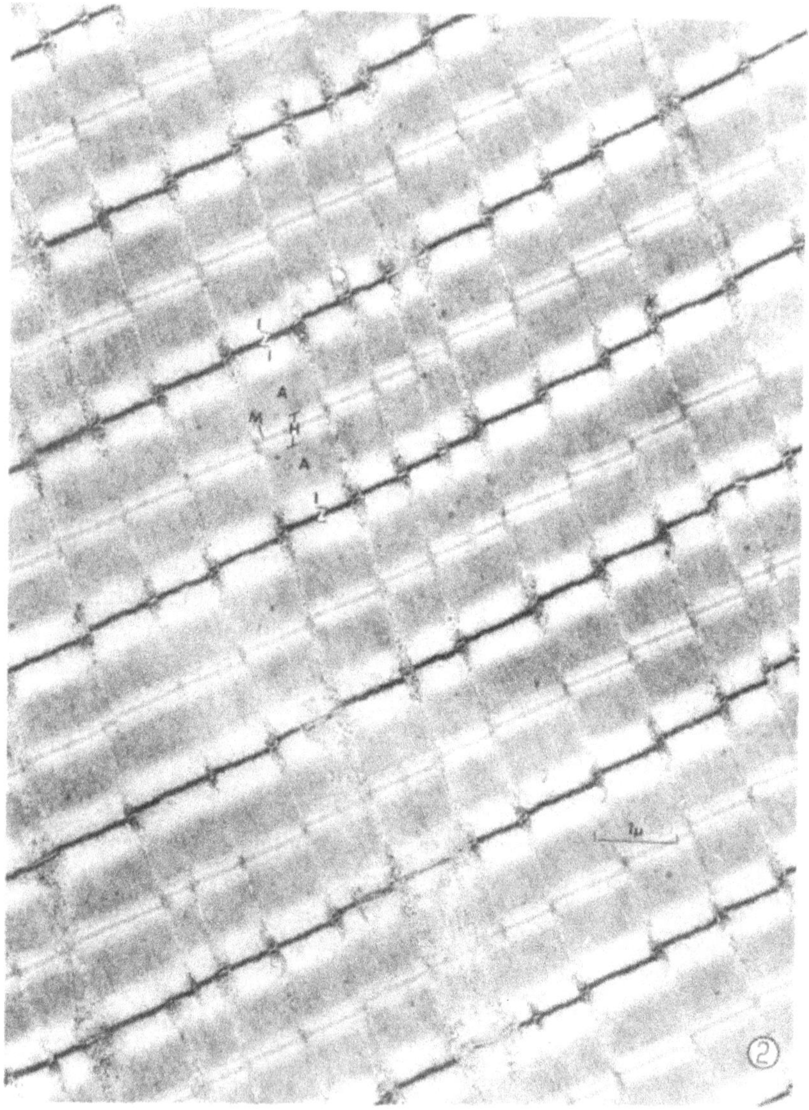

Figure 2. *Longitudinally sectioned sartorius muscle of the frog, Rana pipiens.* The Z lines and the A, I, and H zones of the sarcomeres are indicated. From Smith, 1972.

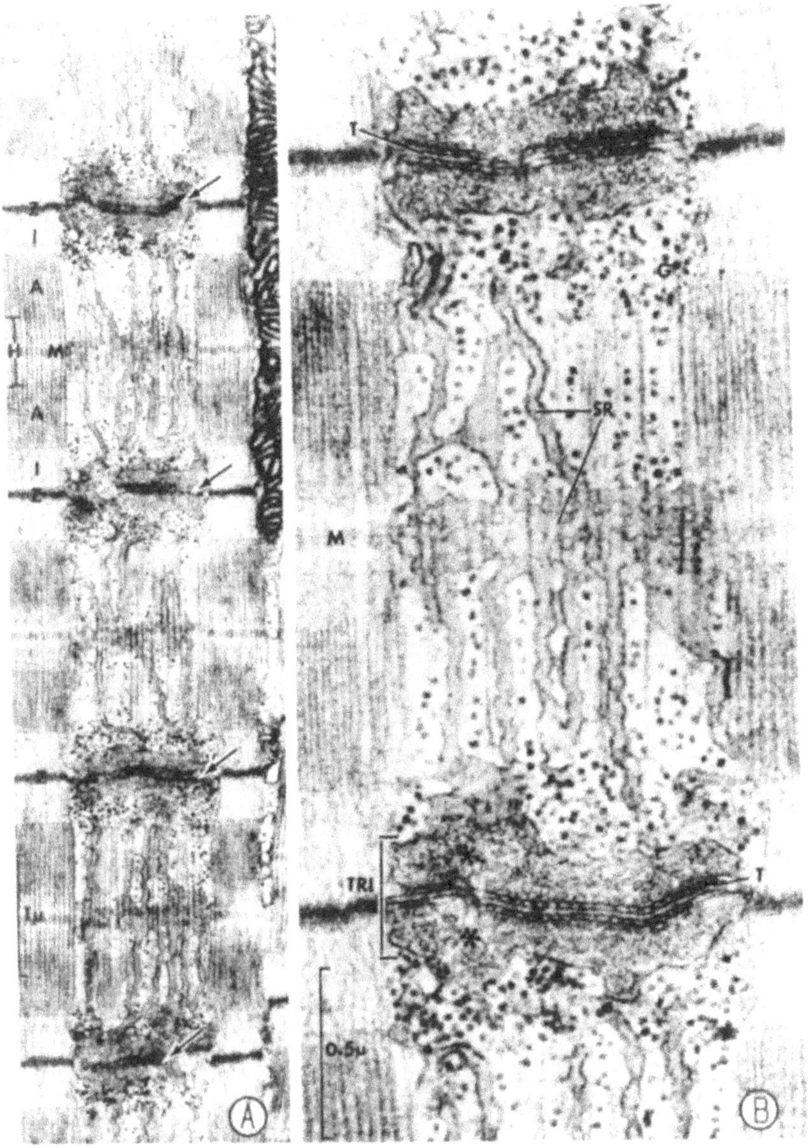

Figure 3. *Longitudinally sectioned sartorius muscle of the frog, Rana pipiens.* The longitudinal tubules of sarcoplasmic reticulum (SR), the terminal cisternae (*), the T-tubules (T), the triad junction (TRI), the Z line (arrow, Z), the A, I, and H zone of the sarcomere, and the M line (M) are indicated. From Smith, 1972.

Table 2. Classification of sarco(endo)plasmic reticulum (SERCA) Ca^{2+}-ATPases

Class	*Tissue*	*Species*	MW	*N Terminal Sequence*	*References*
1 SERCA1a	Adult fast-twitch skeletal muscle	rabbit chicken	109,361 109,375	MEAA MENA	Brandl et al., 1987 Karin et al., 1989
2 SERCA1b	Neonatal fast-twitch skeletal muscle	rabbit	110,331	MEAA	Brandl et al., 1986, 1987
3 SERCA2a	Slow-twitch skeletal, cardiac and smooth muscle	rabbit	109,529	MENA	MacLennan et al., 1985 Brandl et al., 1986, 1987 Lytton and MacLennan, 1988 Lompre et al., 1989
4 SERCA2b	Non-muscle and Smooth muscle	rat	114,759	MENA	Gunteski-Humblin et al., 1988
5 SERCA2b	Non-muscle, and smooth muscle	human	115,000	MENA	Lytton and MacLennan, 1988
6 SERCA2b	Smooth muscle	rabbit	115,000	MENA	Lytton et al., 1989
7 SERCA2b	Smooth muscle	pig	115,000	MENA	Eggermont et al., 1989
8 SERCA3	Both muscle and non-muscle tissues	rat	109,223	MEEA	Burk et al., 1989

(or terminal cisternae) contain electron-dense material in their lumen, consisting of Ca^{2+} binding proteins, such as calsequestrin, that may serve as storage sites for the accumulated Ca^{2+} (Franzini-Armstrong et al., 1987; Lytton and MacLennan, 1992). The slender longitudinal tubules connect the cisternae through the center of the sarcomere and across the Z line; they contain little or no calsequestrin.

Other protein components of the SR include the high affinity Ca^{2+} binding protein, or calreticulin, the histidine-rich Ca^{2+} binding protein, the 50 and 160 kD glycoproteins (sarcalumenins), several proteolipids, and in cardiac sarcoplasmic reticulum the phospholamban (Lytton and MacLennan, 1992; Tada, 1992).

There are significant differences in the protein composition of sarcoplasmic reticulum between species and within each species between muscles of different fiber types. These differences reflect adaptations to unique physiological requirements (Pette and Staron, 1990).

THE CLASSIFICATION OF Ca^{2+}-ATPASE ISOENZYMES

The determination of the amino acid sequences of the sarcoplasmic reticulum Ca^{2+}-ATPase (MacLennan et al., 1985) and of the closely related Na^{+},K^{+}-ATPase (Shull et al., 1985; Kawakami et al., 1985) has opened a new era in the analysis of ion transport mechanisms. Since 1985 several large families of structurally related ion transport enzymes were discovered that are the products of different genes (Table 2). Within each family several isoenzymes may be produced from a single gene product by alternative splicing.

The sarco/endoplasmic reticulum Ca^{2+}-ATPases of mammalian tissues can be divided structurally into three main groups (SERCA1-3) representing the products of different genes.

SERCA1

The SERCA1 gene produces two isoforms of the Ca^{2+}-ATPase that are derived by alternative splicing of the primary gene product. SERCA1a denotes the Ca^{2+}-ATPase of adult fast-twitch skeletal muscle with glycine at its C terminus in the rabbit (Brandl et al., 1987; Lytton et al., 1989), and alanine in the chicken (Campbell et al., 1992). SERCA1b is the alternatively spliced neonatal form of SERCA1, in which the glycine at the C terminus is replaced by the alternative sequence -Asp-Pro-Glu-Asp-Glu-Arg-Arg-Lys (Brandl et al., 1986).

SERCA2

The SERCA2 gene also produces at least two isoforms that are tissue specific. SERCA2a is the principal form of the Ca^{2+}-ATPase in adult slow-twitch skeletal and cardiac muscles and in neonatal skeletal muscles (MacLennan et al., 1985; Brandl et al., 1986, 1987; Lompre et al., 1989). It is also expressed at much lower levels in nonmuscle cells (Burk et al., 1989). SERCA2b is an alternatively spliced product of the same gene. It is located primarily in nonmuscle tissues and in smooth muscles, where it serves as the major intracellular Ca^{2+} pump.

SERCA3

SERCA3 is broadly distributed in skeletal muscle, heart, uterus, and in a variety of nonmuscle cells (Burk et al., 1989). The mRNA levels are particularly high in intestine, lung, and spleen, while it is very low in liver, testes, kidney, and pancreas. In the muscle tissue SERCA3 may be confined primarily to nonmuscle cells (vascular smooth muscle, endothelial cells, etc.).

SERCA Type Ca^{2+}-ATPases From Non-Mammalian Cells

Sequences of SERCA type Ca^{2+}-ATPases were also obtained from *Plasmodium yoelii, Artemia*, and *Drosophila*. These enzymes are similar in size to the SERCA1 and 2a type Ca^{2+}-ATPases from mammalian muscles, but based on their N and C terminal sequences they represent a distinct group.

THE PREDICTED TOPOLOGY OF THE SARCOPLASMIC RETICULUM Ca^{2+}-ATPASES

Combining structural and biochemical information, MacLennan and his colleagues (MacLennan et al., 1985, 1992; Brandl et al., 1986; MacLennan, 1990) constructed a hypothetical model of the tertiary structure of Ca^{2+}-ATPase that has interesting mechanistic implications. The structure was divided into three major parts, designated as the cytoplasmic headpiece, the stalk domain, and the transmembrane domain (Figure 4; MacLennan, 1990), and each was assigned distinct functional and structural roles. Only short loops are assumed to be exposed on the luminal side of the membrane.

The Cytoplasmic Headpiece

The cytoplasmic headpiece contains about 2/3 of the mass of Ca^{2+}-ATPase and consists of five subdomains: the N terminal region (residues 1-40), the transduction or B domain (residues 131-238), the phosphory-

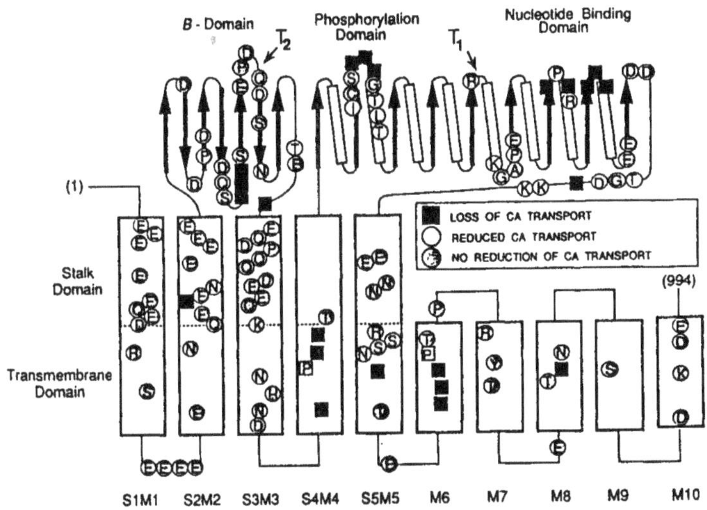

Figure 4. The predicted domain structure of Ca^{2+}-ATPase.
The diagram is based on secondary structure predictions and hydropathy plots. It is divided into three major regions.

1. *The transmembrane domain* consists of 10 helical stretches that form the channel for the translocation of calcium. Mutagenesis of several amino acids in this region interferes with ATP dependent Ca^{2+} transport.
2. *The stalk domain* was introduced to match the electron microscope profile of the Ca^{2+}-ATPase. Mutations in the stalk domain had little effect on Ca^{2+} transport.
3. The cytoplasmic domain is further divided into several subdomains (B-domain, phosphorylation domain, nucleotide binding domain), that contain the active site for ATP hydrolysis (for details see text).

The locations of mutated residues are circled and the wild type residues are identified by a single letter code. The functional consequence of each mutation is indicated by a color code: gray circle = no effect on Ca^{2+} transport; light circle or square = reduced rate of Ca^{2+} transport; black square = background level of Ca^{2+} transport (complete inhibition). Note the clustering of black mutants near the center of the transmembrane domain, where the sites of Ca^{2+} binding are predicted to lie, and in the loops between alpha helices and beta strands in the cytoplasmic domain, where the site of ATP binding is assumed to be. Mutations in the stalk sector and in the periphery of the transmembrane domain had little effect on Ca^{2+} transport function. From MacLennan, 1990.

lation domain (residues 328-505), the nucleotide binding domain (residues 505-680), and the hinge domain (residues 681-738) (Figure 4).

The Phosphorylation and Nucleotide Binding Domains

The phosphorylation domain (residues 328-505) contains the phosphate acceptor Asp351 that is phosphorylated by ATP during Ca^{2+} transport. Site-specific mutagenesis of Asp351 and Lys352 inhibits the phosphorylation of the enzyme by ATP (Maruyama and MacLennan, 1988). The nucleotide binding domain (residues 505-680) contains the binding sites for fluorescein-5′-isothiocyanate (FITC) on Lys515, the site of reaction of adenosine triphosphopyridoxal on Lys684, and the sequences which are homologous with the sites of reaction of 5′-(p-fluorosulfonyl)benzoyl-adenosine and γ[4-(N-2 dichloroethyl-N-methylamino)] benzylamide ATP (ClrATP) in the Na^+,K^+-ATPase (for review see Martonosi, 1992).

The Transduction or B Domain

The transduction or B domain (residues 131-238) lies between two extended helical hairpin loops that traverse the membrane (M_2S_2 and M_3S_3); it consists of seven ß strands separated by short loops that are presumed to fold into an antiparallel ß sandwich (Figure 4; MacLennan, 1990). The B domain is thought to participate in the coupling of ATP hydrolysis to Ca^{2+} translocation.

The Hinge Domain

The proposed hinge region between residues 680 and 738 is one of the most highly conserved segments of the Ca^{2+}-ATPase. By analogy with kinases, MacLennan and his colleagues proposed that the hinge domain interacts with the nucleotide binding and phosphorylation domains transmitting the hinge bending motions of these domains to the Ca^{2+} transport sites (MacLennan et al., 1992).

The Stalk Region

The pentahelical stalk connects the headpiece to the membrane (Figure 4). Its precise role is unknown. In the early models of the enzyme (MacLennan et al., 1985; Brandl et al., 1986) the 18 glutamic acid and three aspartic acid residues in the stalk helices were assumed to form the high affinity binding site for Ca^{2+}, at the entrance to the putative Ca^{2+} transport channel.

However, mutagenesis of the acidic amino acids in the stalk region or in the luminal loop had no effect on ATP-dependent Ca^{2+} transport (Clarke et al., 1989a), while mutagenesis of Glu309, Glu771, Thr799, Asn796, Asp800, and Glu908 in the transmembrane segments inhibited the Ca^{2+}-dependent reactions (Clarke et al., 1989b, 1990a). These observations suggest that the high affinity Ca^{2+} binding site of Ca^{2+}-ATPase is located in the transmembrane helices M_4, M_5, M_6 and M_8, in agreement with energy transfer data (Scott, 1988).

The Transmembrane Domain

The intramembranous part of the ATPase molecule may consist of 8-11 hydrophobic transmembrane helices that anchor the Ca^{2+}-ATPase to the lipid bilayer and form the transmembrane channel for the passage of Ca^{2+}. The intramembranous domain of Ca^{2+}-ATPase contains ~1/3 of the mass of the ATPase molecule. Only small loops are assumed to be exposed on the luminal surface (Matthews et al., 1990; Clarke et al., 1990b; Molnar et al., 1991).

RECONSTRUCTION OF THE THREE-DIMENSIONAL STRUCTURE OF Ca^{2+}-ATPASE BY ELECTRON MICROSCOPY OF NEGATIVELY STAINED Ca^{2+}-ATPASE MEMBRANE CRYSTALS

To understand the molecular mechanism of Ca^{2+} transport we need detailed knowledge about the three-dimensional structure of Ca^{2+}-ATPase at atomic resolution.

During ATP-dependent Ca^{2+} transport the Ca^{2+}-ATPase alternates between two distinct conformations, E_1 and E_2 (Martonosi and Beeler, 1983; Inesi, 1985). The Ca^{2+}-ATPase has been crystallized in both conformations (Dux et al., 1985). The two crystal forms are quite different (Martonosi et al., 1990b), suggesting significant differences between the interactions of Ca^{2+}-ATPase in the E_1 and E_2 conformations. Since the E_1-E_2 transition involves only slight changes in the circular dichroism spectrum of the Ca^{2+}-ATPase, the structural differences between the two states presumably arise by hinge-like or sliding motions of domains rather than by a rearrangement of the secondary structure of the protein (Martonosi et al., 1990a).

The Vanadate-Induced E_2 Type Crystals

The negatively stained vanadate-induced Ca^{2+}-ATPase crystals consist of right-handed helical chains of Ca^{2+}-ATPase dimers (Taylor et al., 1986a,b). Digital Fourier transforms of images of the negatively stained flattened tubules yielded unit cell dimensions of a = 66 Å, b = 114 Å, and $\gamma = 78°$ (Figure 5; Taylor et al., 1984, 1986a,b). The average structure calculated by Fourier synthesis is consistent with ATPase dimers as structural units. The space group of the vanadate-induced E_2-V crystals is p2.

Three-dimensional reconstruction of the structure of vanadate-induced Ca^{2+}-ATPase crystals preserved in uranyl acetate yielded an image of the cytoplasmic region of the molecule (Taylor et al., 1986a,b). In the view normal to the membrane plane, each map shows pear-shaped densities arranged in antiparallel strands that correspond to ribbons of Ca^{2+}-ATPase dimers (Figure 5). The Ca^{2+}ATPase molecules extend about 60 Å above the surface of the bilayer, and their profiles are 65 Å long in a direction parallel to the a axis of the crystal and ~40 Å wide in the direction of the b axis. An 18 Å wide gap separates the two molecules that make up the Ca^{2+}-ATPase dimers; the only visible cytoplasmic connection between them is a 17 Å thick bridge that crosses the intradimer gap at a height of 42 Å above the surface of the bilayer (Taylor et al., 1986a). There are additional interactions in the lipid bilayer. The gap under the bridge is likely to be accessible in the native membrane to ATP, Ca^{2+}, and other cytoplasmic solutes of low molecular weight that interact with the Ca^{2+}-ATPase.

Crystallization of Ca^{2+}-ATPase by Ca^{2+} and Lanthanides in the E_1 State

Analysis of the lanthanide-induced crystalline arrays by negative staining or freeze-fracture electron microscopy reveals obliquely oriented rows of particles, corresponding to individual Ca^{2+}-ATPase molecules (Dux et al., 1985). The unit cell dimensions for the gadolinium-induced Ca^{2+}-ATPase crystals are a = 61.7 Å, b = 54.4 Å, and $\gamma = 111°$ (Figure 6), consistent with a single Ca^{2+}-ATPase monomer per unit cell. The space group of the E_1 type crystals is P1 (Dux et al., 1985).

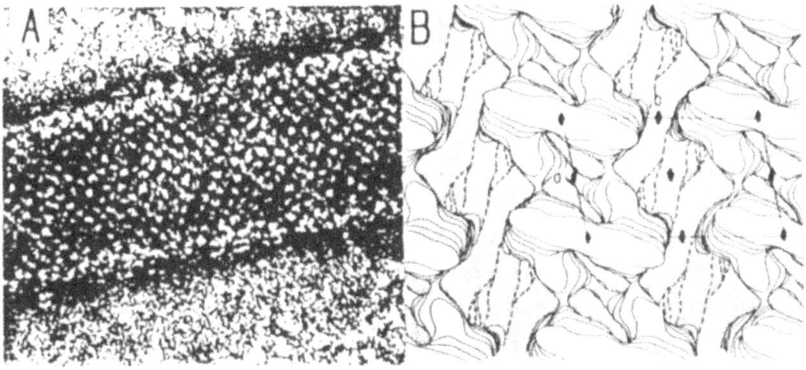

Figure 5. *Vanadate-induced (E2) crystals of Ca^{2+}-ATPase.* (**A**) Negatively stained image of a crystalline tubule. Magnification: 250,000 x. (**B**) Surface contours of Ca^{2+}-ATPase viewed normal to the membrane plane. Stippling indicates the surface of the bilayer and broken lines within the stippled areas are positive density features located within the bilayer. The a and b unit cell axes are marked. From Taylor et al., 1986a.

By proper selection of the experimental conditions a reversible interconversion between the two crystal forms (E_1 and E_2) can be observed, that may be related to the structural transitions between the two major conformations of the Ca^{2+}-ATPase during Ca^{2+} transport.

Crystallization of Ca^{2+}-ATPase in Detergent-Solubilized Sarcoplasmic Reticulum

Microcrystals of Ca^{2+}-ATPase form in detergent-solubilized sarcoplasmic reticulum in the presence of 20 mM Ca^{2+} and 20% glycerol, that contain highly ordered crystalline sheets of Ca^{2+}-ATPase molecules (Figure 7) (Taylor et al., 1988; Stokes and Green, 1990). These represent the first step toward the formation of highly ordered three-dimensional crystals of Ca^{2+} ATPase that may be suitable for X-ray crystallography.

X-RAY AND NEUTRON DIFFRACTION ANALYSIS OF THE Ca^{2+}-ATPASE OF SARCOPLASMIC RETICULUM

In oriented, partially dehydrated multilayers, under conditions suitable for lamellar x-ray diffraction studies, the sarcoplasmic reticulum vesicles retain much of their ATP energized Ca^{2+} transport activity. The

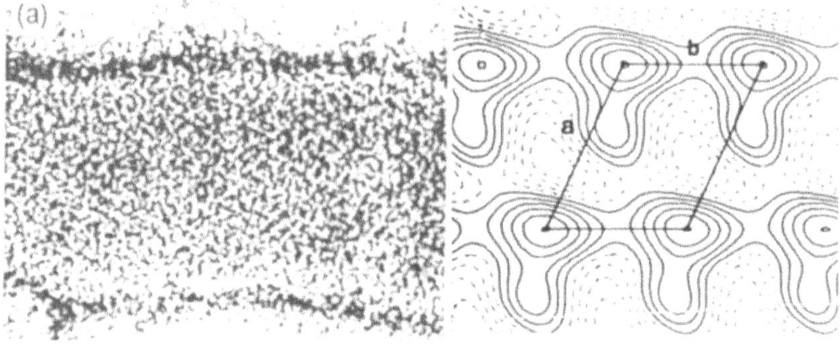

Figure 6. *Ca^{2+}-ATPase crystals induced by praseodymium.* (A) Negatively stained crystalline tubule. Magnification: 276,000 x. (**B**) Projection map of the crystals shown under A with unit cell dimensions marked; a = 66.8 Å, b = 50.4 Å, γ = 114.5°. From Dux et al., 1985.

Ca^{2+} transport can be initiated by flash-photolysis of P^3-1(2-nitro)phenylethyladenosine-5′-triphosphate, "caged ATP." The flash-photolysis of caged ATP rapidly releases ATP and effectively synchronizes the Ca^{2+} transport cycle of the ensemble of Ca^{2+}-ATPase molecules (Blasie et al., 1990).

The changes in the profile structure of the sarcoplasmic reticulum during Ca^{2+} transport imply that about 8% of the total mass of the Ca^{2+}-ATPase is redistributed from the extravesicular surface to the membrane bilayer region and to the intravesicular surface within 200-500 msec after the flash-photolysis of caged ATP (Blasie et al., 1990). During the next 5 seconds there was no further change in the profile structure, although the Ca^{2+}-ATPase completed several cycles of Ca^{2+} transport. This may be explained by assuming that the phosphorylated form of the enzyme is the dominant intermediate during the steady state.

THE MECHANISM OF ACTIVE Ca^{2+} TRANSPORT BY SARCOPLASMIC RETICULUM

The classical studies of Heilbrunn and Wierczinsky (1947), Marsh (1952), Ebashi (1961, 1991), Hasselbach (1964, 1989), and Weber (1966) on the Ca^{2+} regulation of muscle contraction by sarcoplasmic reticulum set the stage for the exploration of the unique role of calcium

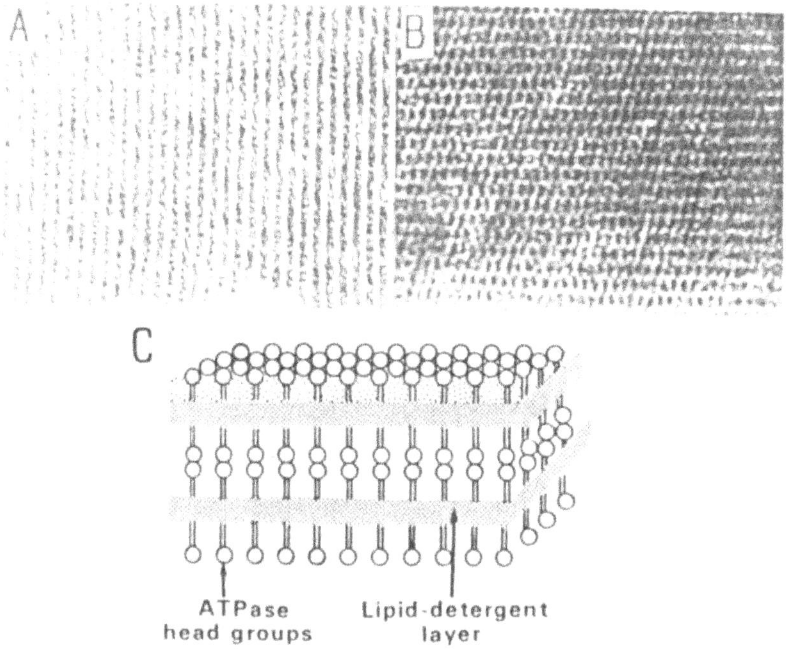

Figure 7. *Three-dimensional Ca^{2+}-ATPase crystals in detergent-solubilized sarcoplasmic reticulum.* Top left (**A**) thin sections; magnification: 207,000x. Top right (**B**) negatively stained lamellae; magnification: 308,000x. Bottom (C) schematic representation of a crystalline lamella. From Taylor et al., 1988.

in the regulation of a wide range of metabolic processes (Berridge, 1988, 1991) and for the elucidation of mechanisms by which the cytoplasmic Ca^{2+} is controlled in living cells (Carafoli, 1987, 1991, 1992).

The Cycle of ATP Hydrolysis

The active transport of calcium from the cytoplasm into the SR lumen occurs at the expense of ATP hydrolysis through the [Mg^{2+} + Ca^{2+}]-activated ATPase. For each ATP hydrolyzed, two Ca^{2+} ions are transferred across the membrane in a series of reactions that are outlined in Figure 8 (Jencks, 1989, 1992; Inesi et al., 1990, 1992a,b).

The Ca^{2+} transport is initiated by the interaction of two Ca^{2+} and one ATP with the Ca^{2+}-ATPase on the cytoplasmic side of the membrane,

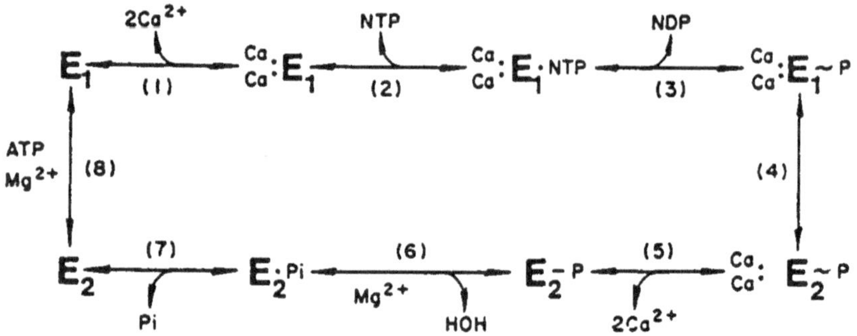

Figure 8. *Elementary steps of ATP hydrolysis and Ca^{2+} transport.* In addition to ATP, other nucleotide triphosphates (NTP) can also serve as energy sources for Ca^{2+} transport releasing the corresponding nucleoside diphosphates (NDP) as the products of the reaction.

The enzyme reacts with Ca^{2+} and NTP in the E_1 conformation (top line) on the cytoplasmic side of the sarcoplasmic reticulum membrane. The change in conformation from the E_1 to the E_2 state (step 4) is associated with the translocation of Ca^{2+} and its release on the luminal side of the membrane (step 5). The reaction is completed by the hydrolysis of $E_2{\sim}P$ yielding inorganic phosphate (Pi) (steps 6-7) and the return of the enzyme to the E_1 conformation ready to start a new cycle (step 8). The two isomerization steps are accelerated by Mg-ATP.

which leads to the formation of an $E_1{<}^{2Ca}_{ATP}$ complex (steps 1 and 2). The binding of the two Ca^{2+} ions is cooperative, with $K_{app} = 2.3 \times 10^6\ M^{-1}$, and a maximum amount of 8 to 10 nmol Ca^{2+}/mg protein can be bound. The binding of Ca^{2+} and ATP induces a conformational change in the enzyme that is detectable by changes in tryptophan fluorescence, and by altered mobility and reactivity of protein-bound spin labels. The ATP is rapidly cleaved, with the phosphorylation of an active site aspartyl residue (step 3), and the occlusion of enzyme-bound Ca^{2+} in a form ($E_1{<}^{\sim P}_{2Ca}$), that is not accessible to EGTA in the medium. The phosphate transfer process (step 3) is reversible, resulting in ATP-ADP exchange. The rate of ATP-ADP exchange may be many times greater than the net rate of ATP hydrolysis.

The phosphorylated enzyme undergoes a conformational change (step 4), resulting in the translocation of Ca^{2+} across the membrane and the conversion of the $E_1{<}^{\sim P}_{2Ca}$ (ADP-sensitive intermediate) into the $E_2{<}^{\sim P}_{2Ca}$ form, that cannot transfer its phosphate to ADP (ADP-insensitive intermediate). Because the $E_2{<}^{\sim P}_{2Ca}$ form has low affinity for

Ca^{2+} (K_m ~1 mM), Ca^{2+} is released into the luminal space of the SR (step 5), where it may accumulate to several millimolar concentration. It is still not settled whether the two Ca^{2+} ions bound to the high affinity site of the Ca^{2+}-ATPase are translocated sequentially (Inesi, 1987) or if there is rapid exchange of the two Ca^{2+} ions after enzyme phosphorylation (Hanel and Jencks, 1991; Orlowski and Champeil, 1991). As the intravesicular Ca^{2+} concentration increases during Ca^{2+} transport, the release of Ca^{2+} from the enzyme is progressively inhibited, with inhibition of ATPase activity and Ca^{2+} transport. The cycle of Ca^{2+} transport is completed by the isomerization and Mg^{2+}-dependent hydrolysis of the phosphoenzyme (steps 6 and 7), with the release of inorganic phosphate on the cytoplasmic surface of the membrane.

The Ca^{2+} pump is capable of lowering the cytoplasmic free Ca^{2+} concentration to or below 10^{-8} M and establish a Ca^{2+} gradient of 1,000-fold, or greater, across the membrane. Much of the accumulated Ca^{2+} is bound in the lumen of the sarcoplasmic reticulum to calsequestrin and other Ca^{2+} binding proteins. With a Ca^{2+} storage capacity of 0.1 μmol Ca^{2+}/mg SR protein and a sarcoplasmic reticulum content of ~5 mg SR protein/g muscle, the sarcoplasmic reticulum is the principal storage site of Ca^{2+} in living muscle.

The sarcoplasmic reticulum membrane contains channels for monovalent cations and anions (Meissner, 1990); the rapid fluxes of these ions through the membrane assures that there are only minor changes in the membrane potential of sarcoplasmic reticulum during the transport and release of Ca^{2+}.

The Ca^{2+} transport process is reversible: in the presence of ADP and inorganic phosphate and at high intravesicular and low external Ca^{2+} concentration, the release of two Ca^{2+} ions across the membrane permits the synthesis of 1 ATP from ADP and inorganic phosphate (Hasselbach, 1978; De Meis, 1981). During reversal of Ca^{2+} transport the Mg^{2+}-enzyme is phosphorylated by inorganic phosphate to yield $E_2{\sim}P_{Mg}$, which is sequentially converted into $E_2{\sim}P_{2Ca}$, and finally $E_1{\sim}P_{2Ca}$. The phosphoryl group is transferred to ADP, forming ATP, with the release of Ca^{2+} into the external medium. The biphasic reaction of the phosphoenzyme intermediate with ADP is consistent with the rapid formation and the slow dissociation of ATP from the active site (Jencks, 1989, 1992).

The reversal of Ca^{2+} transport provided interesting information about the mechanism of the Ca^{2+} pump, but it is not likely to have much physiological relevance because the cellular conditions do not favor reversal.

MUSCLE CONTRACTURE INDUCED BY EXERCISE (BRODY'S DISEASE)

The disorder is characterized by painless muscle stiffness with delayed relaxation provoked by 10-15 seconds of strenuous activity (Brody, 1969). The intolerance to exercise is particularly pronounced at low temperature, creating difficulty of articulation due to stiffness of the tongue and lips. Brief rest permits the muscles to relax fully, after which exercise may be resumed. The slow relaxation of muscle was attributed to a selective defect of the Ca^{2+} uptake function of sarcoplasmic reticulum with elevated levels of sarcoplasmic Ca^{2+} concentration (Brody, 1969).

Immunocytochemical analysis of patients from several families (Karpati et al., 1986; Danon et al., 1988) revealed only trace amounts of Ca^{2+}-ATPase in the fast-twitch type 2 muscle fibers, while the slow-twitch type 1 fibers had the normal amount of Ca^{2+}-ATPase. This suggests that the disease involves selective impairment of the expression of the fast isoenzyme of Ca^{2+}-ATPase in type 2 fibers, while the slow Ca^{2+}-ATPase isoenzyme characteristic of the type 1 fibers is normally produced. The genes for the fast and slow Ca^{2+}-ATPase isoenzymes are located on chromosome 16 and on chromosome 12, respectively (MacLennan et al., 1987). The clinical manifestations of Brody's disease are entirely accountable as the consequence of a deficiency in the SERCA1a isoform of the Ca^{2+}-ATPase in the fast-twitch muscles, while the SERCA2a slow-twitch isoform continues to be expressed.

Continued biochemical and genetic analysis of Brody's disease may shed interesting light on the differential regulation of the concentration of fast and slow Ca^{2+}-ATPase isoenzymes in the sarcoplasmic reticulum. So far there is no reported disease that involves a selective defect in the expression of slow Ca^{2+}-ATPase isoenzyme.

EXCITATION-CONTRACTION (E-C) COUPLING

The Release of Ca^{2+} From Sarcoplasmic Reticulum

Contraction is initiated by the nerve impulse that depolarizes the surface membrane of the muscle cell. The depolarization wave is

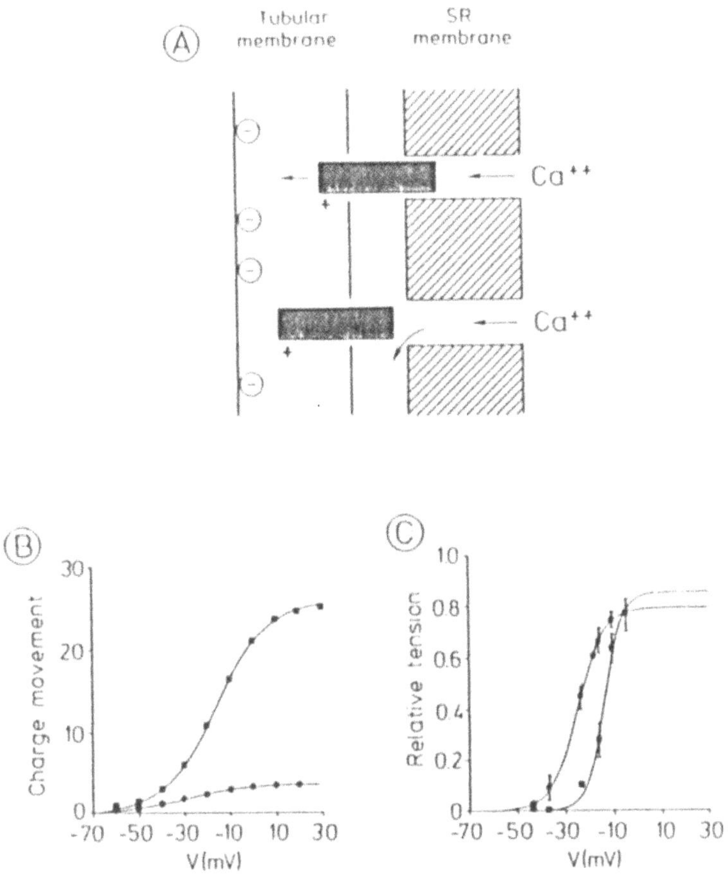

Figure 9. The mechanical hypothesis of the excitation-contraction coupling. (**A**) The depolarization of the T-tubule membrane initiates the movement of a charged structure, presumed to be a segment of the dihydropyridine receptor of the T-tubules, that unplugs the calcium release channel of the sarcoplasmic reticulum causing the release of activating Ca^{2+} into the cytoplasm. (From Ruegg, 1988) **B-C** show the relationship between membrane depolarization and charge movement nC/nF (**B**), or the relative contractile tension (**C**) in rat skeletal muscle. ■, Rat extensor digitorum longus (a fast-twitch muscle). •, Rat soleus (a slow-twitch muscle). In the slow muscle the threshold depolarization is shifted to more negative potential, the charge movement is smaller, and the relationship between charge movement and potential is less steep. These differences are consistent with the smaller density of dihydropyridine receptors in the T-tubules of slow-twitch muscles. From Dulhunty and Gage, 1983.

conducted into the muscle interior by the transverse (T) tubules that establish specialized junctions with the SR. The depolarization of T-tubules triggers a charge movement in the junctional structure (Figure 9) that initiates the series of events culminating in the release of stored calcium from the SR into the cytoplasm (Rios et al., 1991, 1992; Rios and Pizarro, 1991). During contraction the cytoplasmic $[Ca^{2+}]$ increases from resting levels of less than 10^{-7} M to about 10^{-6}-10^{-5} M. Ca^{2+} binding to troponin changes the structure of the thin filaments, permitting the interaction of myosin cross-bridges and F-actin, with activation of the cyclic cleavage of ATP, and development of contractile tension (Ashley et al., 1991).

The relationship between charge movement and Ca^{2+} release is complex and may involve mechanical, electrical, and chemical intermediates (Rios and Pizarro, 1991; Jaimovich, 1991).

One of the chemical intermediates is Ca^{2+}, that may directly or indirectly potentiate Ca^{2+} release from the SR (Rios and Pizarro, 1991). In skeletal muscle the "trigger" Ca^{2+} is not likely to be of extracellular origin as single muscle fibers can twitch in the presence of 80 mM EGTA, that is expected to reduce the external free Ca^{2+} to 10^{-10} M or below (Armstrong et al., 1972; Spiecker et al., 1979). However, in cardiac muscle, extracellular Ca^{2+} plays a significant role in contractile activation (Fabiato, 1989).

The discovery of the second messenger role of inositol 1,4,5-trisphosphate ($InsP_3$) (Berridge, 1988, 1991) has led to speculations about its possible involvement as a messenger in excitation-contraction coupling (Jaimovich, 1991). Mobilization of the Ca^{2+} by $InsP_3$ from intracellular stores has been in fact observed in vascular smooth muscle cells (Somlyo and Somlyo, 1992), but in cardiac and skeletal muscles the role of $InsP_3$ in excitation-contraction coupling is still uncertain (Jaimovich, 1991).

The Structure of T-SR Junctions

The dihydropyridine receptor complex of the T-tubules interacts with the ryanodine-sensitive Ca^{2+} channel of the junctional face membrane of sarcoplasmic reticulum (Figure 10) and triggers the voltage-dependent release of Ca^{2+} from the sarcoplasmic reticulum lumen (Rios and Pizarro, 1991; Ashley et al., 1991; Rios et al., 1991, 1992).

The amino acid sequences of both receptors were deduced from their DNA sequences, and sufficient structural information is now available

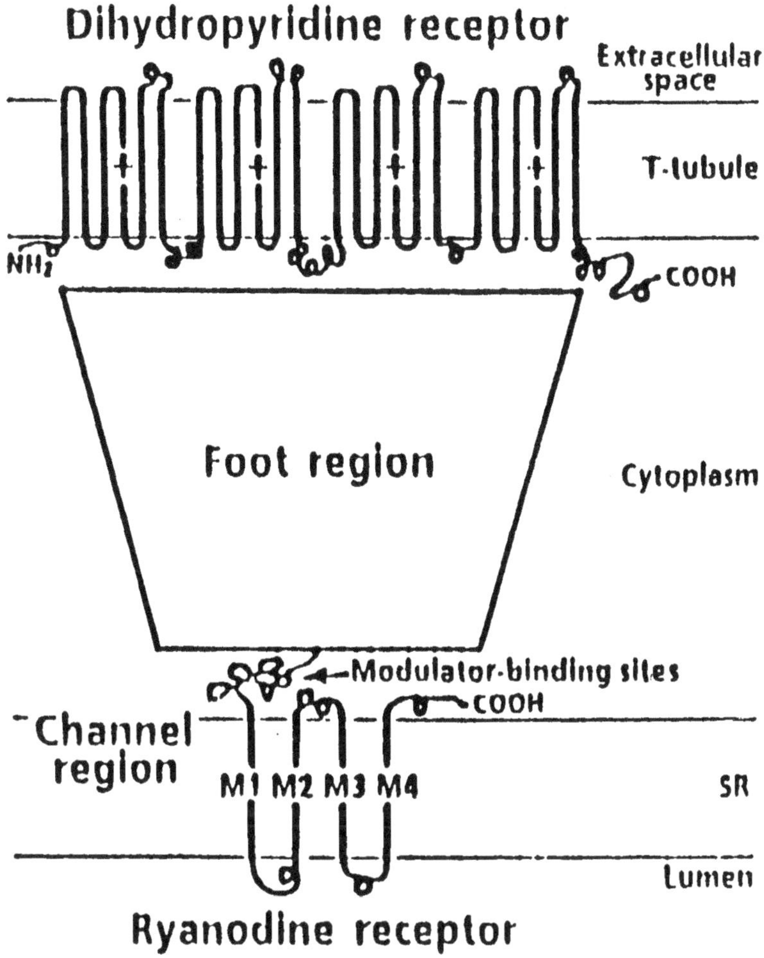

Figure 10. *Schematic representation of the probable interaction of dihydropyridine receptor (DHPR) with the ryanodine receptor (RyR) at the T-tubule-sarcoplasmic reticulum junction.* The $\alpha 1$ subunit of the DHPR contains four homologous regions, each consisting of six transmembrane helices, one of which is positively charged (+). The positively charged helices are presumed to serve as voltage sensors that respond to the depolarization of T-tubule membrane with a change in conformation. The DHPR interacts with the large cytoplasmic domain of the ryanodine receptor that serves as Ca^{2+} release channel. The voltage-dependent change in the conformation of DHPR is transmitted to the RyR, initiating the release of activating calcium from the sarcoplasmic reticulum into the cytoplasm. From Takeshima et al., 1989.

to permit realistic speculation about the molecular mechanism of activating Ca^{2+} release during muscle contraction.

The Dihydropyridine Receptor (DHPR) of Skeletal and Cardiac Muscles

Structural Aspect. The DHP receptors of skeletal and cardiac muscles serve in slightly different ways as voltage sensors for the transmission of excitatory stimulus from the T-tubules to the ryanodine-sensitive Ca^{2+} channel of the sarcoplasmic reticulum (Rios et al., 1991; Rios and Pizarro, 1991; Catterall, 1992). In skeletal muscle the E-C coupling does not require the entry of extracellular Ca^{2+} into the cell and correspondingly less than 5% of the Ca^{2+} channels are active in ion conductance. In cardiac muscle the influx of trigger Ca^{2+} is an essential component of E-Ç coupling, and the cardiac DHP receptor has high Ca^{2+} channel activity and fast response time (Tanabe et al., 1988, 1990a,b, 1991).

The DHP-sensitive Ca^{2+} channels of skeletal and cardiac muscles are coded by different genes (Tanabe et al., 1988, 1990a,b, 1991), and their expression is differentially regulated. In dysgenic mice the functional DHP receptor content is severely reduced in skeletal muscle, causing disruption of excitation-contraction coupling, while in the heart and in the sensory neurons it remains normal (Pincon-Raymond et al., 1985; Beam et al., 1986).

The currently accepted structure of the skeletal muscle DHP receptor complex is that of a pentamer, containing $\alpha1$, $\alpha2$, ß, γ, and δ subunits in 1:1:1:1:1 molar ratio (Catterall, 1992).

The $\alpha1$ subunit is the central component of the DHP receptor. It is present in the purified preparations in two forms with relative molecular masses of 212 and 175 kD. The 175 kD form accounts for more than 90% of the $\alpha1$ subunit content both in purified preparations and in T-tubules. The $\alpha1$ subunit contains binding sites for the three main classes of Ca^{2+} antagonists, the dihydropyridines (nitrendipine, nifedipine, PN 200-110), the phenylalkylamines (verapamil), and the benzothiazepines (diltiazem).

Analysis of the amino acid sequence of the $\alpha1$ subunit reveals four internal structural repeats, each containing five hydrophobic (S1, S2, S3, S5, and S6) and one positively charged (S4) segments (Figure 10). The S1 through S6 segments are assumed to represent α helical membrane spanning domains arranged symmetrically around a central ionic channel, similarly to the structure proposed earlier for the Na^+ channel. The positively charged S4 segments are assumed to serve as voltage sensors. The hydrophilic N- and C-termini are positioned on the cytoplasmic surface.

Muscular Dysgenesis. Muscular dysgenesis (mdg) is a lethal mutation in the mouse, transmitted as a single autosomal recessive trait; it is manifested in a homozygous (mdg/mdg) newborn as a total lack of contractile activity in all skeletal muscles, leading to immediate postnatal death due to inability to breathe.

The defect in dysgenic mouse was traced to the absence or severe deficiency of the dihydropyridine sensitive Ca^{2+} channels (Beam et al., 1986; Romey et al., 1986). Only the $\alpha1$ subunit of the dihydropyridine receptor is affected (Knudson et al., 1989), while the other components of excitation-contraction coupling, including the $\alpha2$, ß, γ, and δ subunits of the dihydropyridine receptor are present in normal concentrations. However, the expression of the $\alpha1$ subunit may be required for the proper targeting of the $\alpha2$ subunit into the T-tubules (Flucher et al., 1991). The excitation-contraction (E-C) coupling was restored by the expression plasmid of the $\alpha1$ subunit of the skeletal muscle DHP receptor in dysgenic cells (Tanabe et al., 1988), providing a good example of the potential of gene therapy.

Differences in the Ca^{2+} Currents of Skeletal, Cardiac, and Dysgenic Skeletal Muscles. In cardiac muscle the action potential rapidly activates the voltage-sensitive Ca^{2+} channel, permitting the entry of Ca^{2+}, that activates further Ca^{2+} release from the sarcoplasmic reticulum by Ca^{2+}-induced Ca^{2+} release (Endo, 1985a,b; Fabiato, 1989). Blocking the Ca^{2+} current by removal of extracellular calcium or by addition of 0.5 mM Cd^{2+} inhibits contraction, confirming the role of extracellular Ca^{2+} in cardiac E-C coupling. Substitution of Ba^{2+} for Ca^{2+} produces a large increase in the L type current of cardiac muscle, while the same substitution causes only a slight change in the Ca^{2+} current of skeletal muscle (Tanabe et al., 1990a,b).

The structural basis of these differences between the skeletal and cardiac muscle DHP receptors was analyzed by expressing their cDNAs in dysgenic myotubes in tissue cultures and studying the characteristics of the spontaneous or electrically evoked contractions (Tanabe et al., 1988, 1990a,b, 1991). The E-C coupling restored by injection of skeletal muscle DHP receptor cDNA (pCAC6) persists in Ca^{2+} free medium or in the presence of 0.5 mM Cd^{2+}, like the E-C coupling of normal skeletal muscle, indicating its independence from extracellular Ca^{2+}. On the other hand, dysgenic muscles injected with the expression plasmid of the coding sequence of cardiac DHP receptor (pCARD1) expressed a Ca^{2+} current (Tanabe et al., 1990a,b, 1991) that was similar to the L type Ca^{2+} current of heart muscle.

Chimeric DHP receptor cDNAs were constructed from the structure of the cardiac DHP receptor, by replacing in the cytoplasmic domain either the N-terminus (CSk1), the I-II loop connecting domains I and II (CSk2), the II-III loop connecting domains II and III (CSk3), the C-terminus (CSk4), or all four of them (CSk7) with the corresponding regions of the skeletal muscle DHP receptor (Tanabe et al., 1990b). All five chimeric cDNAs induced cardiac type rapidly activated Ca^{2+} currents that were evoked at moderate depolarizations and were enhanced by replacing Ca^{2+} with Ba^{2+}. Therefore the characteristics of the cardiac L type channels were retained after the substitution of several cytoplasmic segments with their skeletal muscle counterparts. There were, however, significant differences between the channel constructs in their Ca^{2+} requirement for E-C coupling. The channels induced by chimeras CSk3 and CSk7, like those induced by pCAC6, mediated E-C coupling in the absence of Ca^{2+} or in the presence of 0.5 mM Cd^{2+}, indicating a skeletal muscle type mechanism; under the same conditions the E-C coupling was inhibited in myotubes injected with pCARD1 or CSk1, CSk2, and CSk4 (Tanabe et al., 1990b), consistent with the retention of cardiac characteristics. These observations imply that the II-III loop is the determinant of the skeletal or cardiac type E-C coupling, although other regions, such as the highly conserved III-IV loop, may also contribute. The skeletal or cardiac type activation kinetics of the DHP receptor Ca^{2+} channel is determined by the first of the four internal repeats (Tanabe et al., 1991).

The Ryanodine Receptor of (RyR) the Junctional Sarcoplasmic Reticulum

Purification of the Ryanodine Receptor and its Channel Properties. Several procedures have been introduced for the purification of the ryanodine receptor, all of which yield a protein of high molecular weight (360-560 kD) as the principal component, with ryanodine binding characteristics similar to those of the ryanodine receptor in junctional sarcoplasmic reticulum membrane.

The purified ryanodine receptor induced Ca^{2+} conducting pathways when incorporated into planar bilayers, but the electrical properties and pharmacological responses of the native channels were preserved to different degrees by the different purification procedures. The most detailed comparison with the native sarcoplasmic reticulum was made on the purified ryanodine receptor isolated by immunoaffinity chromatography from triads solubilized by CHAPS and soybean phospholipids (Smith et al., 1986, 1988). The channel kinetics were ligand-dependent and very fast on the microsecond time scale (J. S. Smith et al., 1988), consistent with earlier observations on native SR vesicles. Raising cis-Ca^{2+} from 1 μM to 90 μM increased the open probability (P_0) from 0.014 to 0.85; the Hill coefficient of Ca^{2+} activation (n = 1.33) indicated moderate cooperativity. In the presence of 10 μM cis-Ca^{2+}, addition of 5 and 10 mM ATP to the cis side increased P_0 from 0.16 to 0.60 and 0.93, respectively. Ryanodine (20 nM) after prolonged incubation induced a transition to a state of lower conductance and increased mean open time with occasional channel closures.

The purified receptor channel displayed transitions between several, usually four, conductance states (Smith et al., 1988), that may represent the conductance levels of monomer, dimer, trimer, and tetramer, respectively. These observations are consistent with a tetrameric structure of the ryanodine receptor in which the individual subunits are structurally and functionally linked to form the cooperative high conductance (100 pS) Ca^{2+} release channel. Ryanodine binding to the tetramer may alter the cooperativity between the subunits locking the structure in a dimeric state, with half of the normal conductance and altered gating properties (Smith et al., 1988).

The Amino Acid Sequence of the Skeletal Muscle Ryanodine Receptor. The amino acid sequences of the ryanodine receptors from rabbit (Takeshima et al., 1989; Zorzato et al., 1990) and human skeletal muscles (Zorzato et al., 1990) were deduced from the sequences of their cloned cDNAs. The rabbit ryanodine receptor contains 5037 amino acids with a calculated molecular weight of 565,223 (Takeshima et al., 1989). The cDNA for the human ryanodine receptor encodes a protein of 5032 amino acids with a calculated molecular weight of 563,584 (Zorzato et al., 1990).

The Three Dimensional Architecture of the Ryanodine Receptor. The transmembrane segments of the ryanodine receptors are anchored in the junctional face membrane of the sarcoplasmic reticulum, while the cytoplasmic domains project into the junctional gap forming the feet structures that establish contact with the dihydropyridine receptors on the junctional face of the T-tubules (Figure 10) (Block et al., 1988; Takeshima et al., 1989; MacLennan, 1990; MacLennan and Phillips,

1992). The feet are made up from the cytoplasmic domains of four ryanodine receptor molecules; they form long rows of diamond shaped densities in a tetragonal arrangement touching corner to corner (Block et al., 1988). The size of the subunits is consistent with the molecular weight of the ryanodine receptor monomers.

In tilt view reconstruction of the negatively stained images of isolated noncrystalline feet structures of rabbit skeletal muscle, the receptor complexes appear as square shaped stain-excluding particles with overall dimensions of 27 x 27 nm and a height of 14 nm when viewed normal to the four-fold axis (Wagenknecht et al., 1989). A square-shaped baseplate of 14 x 14 nm area and 4 nm thickness projects from one side of the foot; the baseplate is presumed to represent the membrane embedded channel portion of the ryanodine receptor (Wagenknecht et al., 1989). A similar structure was obtained by cryoelectron microscopy from frozen hydrated specimens (Radermacher et al., 1992).

In the center of the receptor structure there is a 20 nm diameter *center channel* that does not directly extend to either surface. Near the midplane of the structure the center channel branches into four *radial channels* that expand into four stain filled vestibules and open to the cytoplasm on the lateral surfaces of the feet. It is assumed that the gated Ca^{2+} channel is located in the transmembrane domain of the receptor. The channel opens during depolarization of the T-tubules, permitting the release of Ca^{2+} through the center and radial channels into the vestibules and eventually into the cytoplasm.

Malignant Hyperthermia

Malignant hyperthermia (MH) is a catabolic storm (Nelson, 1987, 1988) characterized by rapid increase in body temperature (5°C/hour), muscle contracture, and other signs of metabolic hyperactivity, such as tachycardia, hyperkalemia, hypoxemia, and increased production of carbon dioxide and lactate resulting in acidosis. The attacks are triggered by anesthetic agents, such as halothane and muscle relaxants such as succinylcholine. Once initiated, a vicious cycle develops, in which the metabolic rate increases 10-15-fold above normal and the body temperature may reach 43 °C (109.4 °F). The susceptibility to malignant hyperthermia in humans is inherited through more than one gene or allele and the pattern of inheritance may vary from recessive to dominant (Gronert, 1986). Similar attacks of malignant hyperthermia can be triggered by anesthetics in MH susceptible breeds of pigs and dogs, that serve as useful models for the analysis of the pathogenesis of the disease.

The primary etiological factor is an increase of cytoplasmic [Ca^{2+}] that is provoked by the anesthetics in susceptible individuals. Drugs that increase cytoplasmic Ca^{2+} (cardiac glycosides, lidocaine, caffeine, and Ca salts) enhance, while those that lower cytoplasmic Ca^{2+} (dantrolene, procaine amide) lessen the severity of the hyperthermia attacks. The mortality of the disease decreased from 70% to less than 10% after the introduction of intravenous dantrolene in therapy and prevention (Gronert, 1986).

Localization of the Genetic Defect in the Skeletal Muscle Ryanodine Receptor

There are two isoforms of the ryanodine receptor in mammalian muscles, encoded by two distinct genes. The skeletal muscle ryanodine receptor (RYR1) is expressed in slow- and fast-twitch skeletal muscles (Takeshima et al., 1989; Zorzato et al., 1990; Fujii et al., 1991); its gene is located on human chromosome 19 at 19q13.1 (MacLennan et al., 1990), and on pig chromosome 6 (Harbitz et al., 1990). The cardiac isoform of the ryanodine receptor (RYR2) is expressed only in cardiac muscle (Otsu et al., 1990; Nakai et al., 1990); it shows 66% identity with RYR1 and its gene is located on human chromosome 1 (Otsu et al., 1990).

The RYR1 gene is at least 240 kb in length and contains close to 100 exons (MacLennan and Phillips, 1992). A variety of polymorphisms have been identified, most of which are not related to malignant hyperthermia susceptibility.

In a study of five breeds of pigs (Pietrain, Yorkshire, Poland China, Duroc, and Landrace), that included a total of 376 animals, complete linkage was observed between malignant hyperthermia susceptibility (MHS) and the mutation of C1843 to T1843 (CGC $\rightarrow$ TGC), in the nucleotide sequence of full length pig RYR1 cDNA; this resulted in the substitution of Arg 615 to Cys 615 in the ryanodine receptor (Fujii et al., 1991; Otsu et al., 1991; MacLennan and Phillips, 1992).

The substitution of Cys for Arg facilitates the opening and inhibits the closing of the RYR1 Ca^{2+} channel by altering its affinities for the various regulatory ligands (Ca^{2+}, Mg^{2+}, ATP, calmodulin, halothane, ryanodine, etc.).

In humans, malignant hyperthermia susceptibility is inherited through more than one gene (Gronert, 1986; MacLennan and Phillips, 1992), and in addition to changes in the ryanodine receptor (MacLennan et al., 1990; Gillard et al., 1991), other types of genetic defects involving fatty acids (Fletcher et al., 1990) or the inositol-1,4,5-trisphosphate system (Foster, 1990) may also contribute to the malignant hyperthermia susceptibility.

SUMMARY

Sarcoplasmic reticulum is the principal regulator of cytoplasmic free Ca^{2+} concentration in muscle.

The Ca^{2+} stimulated ATPase enzyme of sarcoplasmic reticulum transports Ca^{2+} from the cytoplasm into the sarcoplasmic reticulum lumen and maintains the low cytoplasmic Ca^{2+} concentration during muscle relaxation.

The muscle contraction is initiated by the nerve impulse that travels through the surface membrane and T-tubules to the sarcoplasmic reticulum. The action potential triggers a conformational change (charge movement) in the voltage-sensitive dihydropyridine receptors of the T-tubules. This structural change in the dihydropyridine receptor activates the ryanodine sensitive Ca^{2+} channels in the junctional membrane of sarcoplasmic reticulum and initiates the rapid release of Ca^{2+} from the sarcoplasmic reticulum into the cytoplasm. Saturation of troponin C with Ca^{2+} triggers muscle shortening and tension development.

Genetic defect in the voltage sensitive dihydropyridine receptor causes dysgenesis, mutation in the ryanodine receptor leads to malignant hyperthermia, while impaired expression of the fast-twitch isoform of the Ca^{2+}-ATPase slows the relaxation of muscle.

REFERENCES

Armstrong, C.M., Bezanilla, F.M., & Horowicz, P. (1972). Twitches in the presence of ethylene glycol bis (β- aminoethyl ether) - N, N^1 - tetraacetic acid. Biochim. Biophys. Acta. 267, 605-608.

Ashley, C. C., Mulligan, I. P., & Lea, T. J. (1991). Ca^{2+} and activation mechanisms in skeletal muscle. Quart. Rev. Biophys. 24, 1-73.

Beam, K. G., Knudson, C. M., & Powell, J. A. (1986). A lethal mutation in mice eliminates the slow calcium current in skeletal muscle cells. Nature 320, 168-170.

Berridge, M. J. (1988). The Croonian lecture, 1988. Inositol lipids and calcium signalling. Proc. Roy. Soc. Lond. B 234, 359-378.

Berridge, M. J. (1991). Cytoplasmic calcium oscillations: a two pool model. Cell Calcium 12, 63-72.

Blasie, J. K., Pascolini, D., Asturias, F., Herbette, L. G., Pierce, D., & Scarpa, A. (1990). Large scale structural changes in the sarcoplasmic reticulum ATPase appear essential for calcium transport, Biophys. J. 58, 687-693.

Block, B. A., Imagawa, T., Campbell, K. P., & Franzini-Armstrong, C. (1988). Structural evidence for direct interaction between the molecular components of the transverse tubule/sarcoplasmic reticulum junction in skeletal muscle. J. Cell Biol. 107, 2587-2600.

Brandl, C. J., Green, N. M., Korczak, B., & MacLennan, D. H. (1986). Two Ca^{2+}-ATPase genes: homologies and mechanistic implications of deduced amino acid sequences, Cell 44, 597-607.

Brandl, C. J., deLeon, S., Martin, D. R., & MacLennan, D. H. (1987). Adult forms of the Ca^{2+} ATPase of sarcoplasmic reticulum. Expression in developing skeletal muscle. J. Biol. Chem. 262, 3768-3774.

Brody, I. A. (1969). Muscle contracture induced by exercise. A syndrome attributable to decreased relaxing factor. New England J. Med. 281, 187-192.

Burk, S. E., Lytton, J., MacLennan, D. H., & Shull, G. E. (1989). cDNA cloning, functional expression and mRNA tissue distribution of a third organellar Ca^{2+} pump. J. Biol. Chem. 264, 18561-18568.

Campbell, A. M., Kessler, P. D., & Fambrough, D. M., (1992). The alternative carboxyl termini of avian cardiac and brain sarcoplasmic reticulum endoplasmic reticulum Ca^{2+}-ATPases are on opposite sides of the membrane. J. Biol. Chem. 267, 9321-9325.

Carafoli, E. (1987). Intracellular calcium homeostasis. Ann. Rev. Biochem. 56, 395-433.

Carafoli, E. (1988). Membrane transport of calcium: an overview. Methods in Enzymology 157, 3-11.

Carafoli, E. (1991). Calcium pump of the plasma membrane. Physiol. Rev. 71, 129-153.

Carafoli, E. (1992). The Ca^{2+} pump of the plasma membrane. J. Biol. Chem. 267, 2115-2118.

Catterall, W. A. (1992). Molecular properties of voltage-gated ion channels in the heart. In: The Heart and Cardiovascular System: Scientific Foundations, 2nd ed., Volume 1 (Fozzard, H.A., Jennings, R.B., Haber, E., Katz, A.M., & Morgan, H.E., (Eds.), pp. 945-962, Raven Press, New York.

Clarke, D. M., Maruyama, K., Loo, T. W., Leberer, E., Inesi, G., & MacLennan, D. H. (1989a). Functional consequences of glutamate, aspartate, glutamine, and asparagine mutations in the stalk sector of the Ca^{2+}-ATPase of sarcoplasmic reticulum. J. Biol. Chem. 264, 11246-11251.

Clarke, D. M., Loo, T. W., Inesi, G., & MacLennan, D. H. (1989b). Location of high affinity Ca^{2+}-binding sites within the predicted transmembrane domain of the sarcoplasmic reticulum Ca^{2+}-ATPase. Nature 339, 476-478.

Clarke, D. M., Loo, T. W., & MacLennan, D. H. (1990a). Functional consequences of alterations to polar amino acids located in the transmembrane domain of the Ca^{2+}-ATPase of sarcoplasmic reticulum. J. Biol. Chem. 265, 6262-6267.

Clarke, D. M., Loo, T. W., & MacLennan, D. H. (1990b). The epitope for monoclonal antibody A20 (amino acids 870-890) is located on the luminal surface of the Ca^{2+}-ATPase of sarcoplasmic reticulum. J. Biol. Chem. 265, 17405-17408.

Danon, M. J., Karpati, G., Charuk, J., & Holland, P. (1988). Sarcoplasmic reticulum adenosine triphosphatase deficiency with probable autosomal dominant inheritance. Neurology 38, 812-815.

De Meis, L. (1981). The Sarcoplasmic Reticulum. Transport and Energy Transduction. Wiley Interscience, New York.

Dulhunty, A. F. & Gage, P. W. (1983). J. Physiol. (London) 341, 213-231.

Dux, L., Taylor, K. A., Ting-Beall, H. P., & Martonosi, A. (1985). Crystallization of the Ca^{2+}-ATPase of sarcoplasmic reticulum by calcium and lanthanide ions. J. Biol. Chem. 260, 11730-11743.

Ebashi, S. (1961). Calcium binding activity of vesicular relaxing factor. J. Biochem. (Tokyo) 50, 236-244.

Ebashi, S. (1991). Excitation-contraction coupling and the mechanism of muscle contraction. Ann. Rev. Physiol. 53, 1-16.

Eggermont, J. A., Wuytack, F., De Jaegere, S., Nelles, L. & Casteels, R. (1989). Biochem. J. 260, 757-761.

Endo, M. (1985a). Calcium release from sarcoplasmic reticulum. Curr. Top. Membr. Transp. 25, 181-230.

Endo, M. (1985b). The calcium - induced calcium release from the sarcoplasmic reticulum. In: Structure and Function of Sarcoplasmic Reticulum, Symposium on Structure and Function of Sarcoplasmic Reticulum, Kobe, Japan, November 1-4, 1982 (Fleisher, S., & Tonomura, Y., Eds.), pp. 521-535, Academic Press, Orlando.

Fabiato, A. (1989). Appraisal of the physiological relevance of two hypotheses for the mechanism of calcium release from the mammalian cardiac sarcoplasmic reticulum: calcium-induced release versus charge-coupled release. Mol. Cell. Biochem. 89, 135-140.

Fletcher, J. E., Tripolitis, L., Erwin, K., Hanson, S., Rosenberg, H., Conti, P. A., & Beech, J. (1990). Fatty acids modulate calcium-induced calcium release from skeletal muscle heavy sarcoplasmic reticulum fractions: implications for malignant hyperthermia. Biochem. Cell Biol. 68, 1195-1201.

Flucher, B. E., Phillips, J. L., & Powell, J. A. (1991). Dihydropyridine receptor α subunits in normal and dysgenic muscle in vitro: expression of α_1 is required for proper targeting and distribution of α_2. J. Cell Biol. 115, 1345-1356.

Foster, P. S. (1990). Malignant hyperpyrexia. Int. J. Biochem. 22, 1217-1222.

Franzini-Armstrong, C., Kenney, L. J., & Varriano-Marston, E. (1987). The structure of calsequestrin in triads of vertebrate skeletal muscle: a deep-etch study. J. Cell Biol. 105, 49-56.

Fujii, J., Otsu, K., Zorzato, F., de Leon, S., Khanna, V. K., Weiler, J. E., O'Brien, P. J., & MacLennan, D. H. (1991). Identification of a mutation in porcine ryanodine receptor associated with malignant hyperthermia. Science 253, 448-451.

Gillard, E. F., Otsu, K., Fujii, J., Khanna, V. K., de Leon, S., Derdemezi, J., Britt, B. A., Duff, C. L., Worton, R. G., & MacLennan, D. H. (1991). A substitution of cysteine for arginine 614 in the ryanodine receptor is potentially causative of human malignant hyperthermia. Genomics 11, 751-755.

Gronert, G. A. (1986). Malignant hyperthermia In: Myology, Vol. 2 (Engel, A. G., & Banker, B. Q., eds.), pp. 1763-1784, McGraw Hill, New York.

Gunteski-Hamblin, A - M., Greeb, J., & Shull, G. E. (1988). A novel Ca^{2+} pump expressed in brain, kidney, and stomach is encoded by an alternative transcript of the slow-twitch muscle sarcoplasmic reticulum Ca - ATPase gene. Identification of CDNAs encoding CA^{2+} and other cation - transporting ATPases using an oligonucleotide probe derived from the ATP - binding site. J. Biol. Chem. 263, 15032-15040.

Hanel A. M., & Jencks, W. P. (1991). Dissociation of calcium from the phosphorylated calcium-transporting adenosine triphosphatase of sarcoplasmic reticulum: kinetic equivalence of the calcium ions bound to the phosphorylated enzyme. Biochemistry 30, 11320-11330.

Harbitz, I., Chowdhary, B., Thomsen, P. D., Davies, W., Kaufmann, U., Kran, S., Gustavsson, I., Christensen, K., & Hauge, J. G. (1990). Assignment of the porcine calcium release channel gene, a candidate for the malignant hyperthermia locus, to the 6p11----q21 segment of chromosome 6. Genomics 8, 243-248.

Hasselbach, W. (1964). Relaxing factor and the relaxation of muscle. Prog. Biophys. Mol. Biol. 14, 167-222.

Hasselbach, W. (1978). The reversibility of the sarcoplasmic calcium pump. Biochim. Biophys. Acta 515, 23-53.

Hasselbach, W. (1989). From frog lung to calcium pump. In: Membrane Transport. People and Ideas (Tosteson, D. C., Ed.), pp. 187-201, American Physiological Society, Bethesda, MD.

Heilbrunn, L. V., & Wiercinsky, F. J. (1947). Action of various cations on muscle protoplasm. J. Cell. Comp. Physiol. 19, 15-32.

Inesi, G. (1985). Mechanism of calcium transport. Ann. Rev. Physiol. 47, 573-601.

Inesi, G. (1987). Sequential mechanism of calcium binding and translocation in sarcoplasmic reticulum adenosine triphosphatase. J. Biol. Chem. 262, 16338-16342.

Inesi, G., Sumbilla, C., & Kirtley, M. E. (1990). Relationships of molecular structure and function in Ca^{2+}-transport ATPase. Physiol. Rev. 70, 749-760.

Inesi, G., Lewis, D., Nikic, D., Hussain, A., & Kirtley, M. E. (1992a). Long-range intramolecular linked functions in the calcium transport ATPase. Adv. Enzymol. & Related Areas of Molec. Biol. 65, 185-215.

Inesi, G., Cantilina, T., Yu, X., Nikic, D., Sagara, Y., & Kirtley, M. E. (1992b). Long-range intramolecular linked functions in activation and inhibition of SERCA ATPases. Ann. NY Acad. Sci. 671, 32-48.

Jaimovich, E. (1991). Chemical transmission at the triad: $InsP_3$? J. Musc. Res. Cell Motil. 12, 316-320.

Jencks, W. P. (1989). How does a calcium pump pump calcium? J. Biol. Chem. 264, 18855-18858.

Jencks, W. P. (1992). On the mechanism of ATP-driven Ca^{2+} transport by the calcium ATPase of sarcoplasmic reticulum. Ann. NY Acad. Sci. 671, 49-57.

Karin, N. J., Kaprielian, Z., & Fambrough, D. M. (1989). Expression of avian Ca^{2+}-ATPase in cultured mouse myogenic cells. Mol. Cell Biol. 9, 1978-1986.

Karpati, G., Charuk, J., Carpenter, S., Jablecki, C., & Holland, P. (1986). Myopathy caused by a deficiency of Ca^{2+}-adenosine triphosphatase in sarcoplasmic reticulum (Brody's disease). Ann. Neurol. 20, 38-49.

Kawakami, K., Noguchi, S., Noda, M., Takahashi, H., Ohta, T., Kawamura, M., Nojima, H., Nagano, K., Hirose, T., Inayama, S., Hayashida, H., Miyata, T., & Numa, S. (1985). Primary structure of the α-subunit of Torpedo californica (Na^+ + K^+)ATPase deduced from cDNA sequence. Nature (London) 316, 733-736.

Knudson, C. M., Chaudhari, N., Sharp, A. H., Powell, J. A., Beam, K. G., & Campbell, K. P. (1989). Specific absence of the $_{-1}$ subunit of the dihydropyridine receptor in mice with muscular dysgenesis. J. Biol. Chem. 264, 1345-1348.

Loewenstein, W. R. (1989). From cell theory to cell connectivity: Experiments in cell-to-cell communication. In: Membrane Transport. People and Ideas (Tosteson, D. C., ed.), pp. 303-335, American Physiological Society, Bethesda, MD.

Lompre, A.M., de la Bastie, D., Boheler, K. R., & Schwartz, K. (1989). Characterization and expression of the rat heart sarcoplasmic reticulum Ca^{2+}-ATPase mRNA. FEBS Lett. 249, 35-41.

Lytton, J., & MacLennan, D. H. (1988). Molecular cloning of CDNAs from human kidney coding for two alternatively spliced products of the cardiac Ca^{2+}-ATPase gene. J. Biol. Chem. 263, 15024-15031.

Lytton, J., & MacLennan, D. H. (1992). Sarcoplasmic reticulum. In: The Heart and Cardiovascular System: Scientific Foundations, 2nd ed., Vol. 2 (Fozzard, H. A., Jennings, R. B., Haber, E., Katz, A. M., & Morgan, H. E., Eds.), pp. 1203-1222, Raven Press, New York.

Lytton, J., Zarain-Herzberg, A., Periasamy, M., & MacLennan, D. H. (1989). Molecular cloning of the mammalian smooth muscle sarco(endo)plasmic reticulum Ca^{2+}-ATPase. J. Biol. Chem. 264, 7059-7065.

MacLennan, D. H. (1990). Molecular tools to elucidate problems in excitation-contraction coupling. Biophys. J. 58, 1355-1365.

MacLennan, D. H., & Phillips, M. S. (1992). Malignant hyperthermia. Science 256, 789-794.

MacLennan, D. H., Brandl, C. J., Korczak, B., & Green, N. M. (1985). Amino-acid sequence of a Ca^{2+} + Mg^{2+}-dependent ATPase from rabbit muscle sarcoplasmic reticulum, deduced from its complementary DNA sequence. Nature (London) 316, 696-700.

MacLennan, D. H., Brandl, C. J., Champaneria, S., Holland, P. C., Powers, V. E., & Willard, H. F. (1987). Fast-twitch and slow-twitch/cardiac Ca^{2+} ATPase genes map to human chromosomes 16 and 12. Somat. Cell Molec. Genet. 13, 341-346.

MacLennan, D. H., Duff, C., Zorzato, F., Fujii, J., Phillips, M., Korneluk, R. G., Frodis, W., Britt, B. A., & Worton, R. G. (1990). Ryanodine receptor gene is a candidate for predisposition to malignant hyperthermia. Nature (London) 343, 559-561.

MacLennan, D. H., Toyofuku, T., & Lytton, J. (1992). Structure-function relationships in sarcoplasmic or endoplasmic reticulum type Ca^{2+} pumps. Ann. NY Acad. Sci. 671, 1-10.

Marsh, B. B. (1952). The effects of adenosine triphosphate on the fibre volume of a muscle homogenate. Biochim. Biophys. Acta 9, 247-260.

Martonosi, A. N. (1983). The regulation of cytoplasmic Ca^{2+} in muscle and nonmuscle cells. In: Muscle and Nonmuscle Motility, Volume 1, (Stracher, A., ed.), pp. 233-357, Academic Press, New York.

Martonosi, A. (1992). The Ca^{2+} transport ATPases of sarco(endo) plasmic reticulum and plasma membranes. In: Molecular Aspects of Transport Proteins (de Pont, J. J. H. H. M., ed.), pp. 57-116, Elsevier, Amsterdam.

Martonosi, A. N., & Beeler, T. J. (1983). Mechanism of Ca^{2+} transport by sarcoplasmic reticulum. In: Handbook of Physiology. Section 10, Skeletal Muscle, Ch. 15 (Peachey, L. D., and Adrian, R. H., eds.), pp. 417-485, American Physiological Society, Bethesda, MD.

Martonosi, A. N., Jona, I., Molnar, E., Seidler, N. W., Buchet, R., & Varga, S. (1990a). Emerging views on the structure and dynamics of the Ca^{2+}-ATPase in sarcoplasmic reticulum. FEBS Lett. 268, 365-370.

Martonosi, A., Taylor, K. A., & Pikula, S. (1990b). The crystallization of the Ca^{2+}-ATPase of sarcoplasmic reticulum. In: Crystallization of Membrane Proteins (Michel, H., ed.), pp. 167-182, CRC Press, Boca Raton, FL.

Maruyama, K. & MacLennan, D. H. (1988). Mutation of aspartic acid-351, lysine-352, and lysine-515 alters the Ca^{2+} transport activity of the Ca^{2+}-ATPase expressed in COS-1 cells. Proc. Natl. Acad. Sci. USA 85, 3314-3318.

Matthews, I., Sharma, R. P., Lee, A. G., & East, J. M. (1990). Transmembranous organization of (Ca^{2+}-Mg^{2+})-ATPase from sarcoplasmic reticulum. Evidence for lumenal location of residues 877-888. J. Biol. Chem. 265, 18737-18740.

Meissner, G. (1990). Calcium release and monovalent ion channels in sarcoplasmic reticulum. In: Cardiac Electrophysiology. From Cell to Bedside (Zipes, D. P., & Jalife, J., Eds.), pp. 70-76, W. B. Saunders Co., Philadelphia, PA.

Molnar, E., Varga, S., Jona, I., & Martonosi, A. (1991). Covalent labeling of the cytoplasmic or luminal domains of the sarcoplasmic reticulum Ca^{2+}-ATPase with fluorescent azido dyes. Biochim. Biophys. Acta 1068, 27-40.

Nakai, J., Imagawa, T., Hakamata, Y., Shigekawa, M., Takeshima, H., & Numa, S. (1990). Primary structure and functional expression from cDNA of the cardiac ryanodine receptor/calcium release channel. FEBS Lett. 271, 169-177.

Nelson, T. E. (1987). Skeletal muscle sarcoplasmic reticulum in the malignant hyperthermia syndrome. In: Malignant Hyperthermia (Britt, B. A., ed.), pp. 43-78, Martinus Nijhoff Publ., Boston, MA.

Nelson, T. E. (1988). SR function in malignant hyperthermia. Cell Calcium 9, 257-265.

Orlowski, S. & Champeil, P. (1991). The two calcium ions initially bound to nonphosphorylated sarcoplasmic reticulum Ca^{2+}-ATPase can no longer be kinetically distinguished when they dissociate from phosphorylated ATPase toward the lumen. Biochemistry 30, 11331-11342.

Otsu, K., Willard, H. F., Khanna, V. K., Zorzato, F., Green, N. M., & MacLennan, D. H. (1990). Molecular cloning of cDNA encoding the Ca^{2+} release channel (ryanodine receptor) of rabbit cardiac muscle sarcoplasmic reticulum. J. Biol. Chem. 265, 13472-13483.

Otsu, K., Khanna, V. K., Archibald, A. L., & MacLennan, D. H. (1991). Cosegregation of porcine malignant hyperthermia and a probable causal mutation in the skeletal muscle ryanodine receptor gene in backcross families. Genomics 11, 744-750.

Peachey, L. D. & Franzini-Armstrong, C. (1983). Structure and function of membrane systems of skeletal muscle cell. In: Handbook of Physiology. Section 10, Skeletal Muscle (Peachey, L. D., Adrian, R. H., & Geiger, S. R., eds.), pp. 23-71, American Physiological Society, Bethesda, MD.

Pette, D. & Staron, R. S. (1990). Cellular and molecular diversities of mammalian skeletal muscle fibers. Rev. Physiol. Biochem. Pharmacol. 116, 1-76.

Pincon-Raymond, M., Rieger, F., Fosset, M., & Lazdunski, M. (1985). Abnormal transverse tubule system and abnormal amount of receptors for Ca^{2+} channel inhibitors of the dihydropyridine family in skeletal muscle from mice with embryonic muscular dysgenesis. Devel. Biol. 112, 458-466.

Radermacher, M., Wagenknecht, T., Grassucci, R., Frank, J., Inui, M., Chadwick, C., & Fleischer, S. (1992). Cryo-EM of the native structure of the calcium release channel/ryanodine receptor from sarcoplasmic reticulum. Biophys. J. 61, 936-940.

Rios, E. & Pizarro, G. (1991). Voltage sensor of excitation-contraction coupling in skeletal muscle. Physiol. Rev. 71, 849-908.

Rios, E., Ma, J. J., & Gonzalez, A. (1991). The mechanical hypothesis of excitation-contraction (EC) coupling in skeletal muscle. J. Musc. Res. Cell Motil. 12, 127-135.

Rios, E. Pizarro, G., & Stefani, E. (1992). Charge movement and the nature of signal transduction in skeletal muscle excitation-contraction coupling. Ann. Rev. Physiol. 54, 109-133.

Romey, G., Rieger, F., Renaud, J. F., Pincon-Raymond, M., & Lazdunski, M. (1986). The electrophysiological expression of Ca^{2+} channels and of apamin sensitive Ca^{2+} activated K^{+} channels is abolished in skeletal muscle cells from mice with muscular dysgenesis. Biochem. Biophys. Res. Commun. 136, 935-940.

Ruegg, J. C. (1988). Calcium in Muscle Activation, Springer Verlag, Berlin.

Scarpa, A., Carafoli, E., & Papa, S., Eds. (1992). Ion-motive ATPases: Structure, Function and Regulation. Ann. NY Acad. Sci. Vol. 671, NY Acad. Sci., New York.

Scott, T. L. (1988). Molecular topography of the Ca^{2+}-ATPase of sarcoplasmic reticulum. Mol. Cell. Biochem. 82, 51-54.

Shull, G. E., Schwartz, A., & Lingrel, J. B. (1985). Amino-acid sequence of the catalytic subunit of the (Na^{+} + K^{+})-ATPase deduced from a complementary DNA. Nature 316, 691-695.

Smith, D. S. (1972). Muscle. Academic Press, New York.

Smith, J. S., Coronado, R., & Meissner, G. (1986). Single channel measurements of the calcium release channel from skeletal muscle sarcoplasmic reticulum. Activation by Ca^{2+} and ATP and modulation by Mg^{2+}. J. Gen. Physiol. 88, 573-588.

Smith, J. S., Imagawa, T., Ma, J., Fill, M., Campbell, K. P., & Coronado, R. (1988). Purified ryanodine receptor from rabbit skeletal muscle is the calcium-release channel of sarcoplasmic reticulum. J. Gen. Physiol. 92, 1-26.

Somlyo, A. P. & Somlyo, A. V. (1992). Smooth muscle structure and function. In: The Heart and Cardiovascular System, Vol. 2, 2nd ed. (Fozzard, H. A., Jennings, R. B., Haber, E., Katz, A. M., & Morgan, H. E., eds.), pp. 1295-1324, Raven Press, Ltd., New York.

Spiecker, W., Melzer, W., & Luttgau, H. C. (1979). Extracellular Ca^{2+} and excitation-contraction coupling. Nature 280, 158-160.

Stein, W. D. (ed.), (1988). The Ion Pumps: Structure, Function and Regulation, Shoresh Workshop on Ion Pumps, Shoresh, Israel, August 30-September 2, 1987, Alan R. Liss, Inc., New York.

Stein, W. D. (1990). Channels, Carriers and Pumps. An Introduction to Membrane Transport. Academic Press, New York.

Stokes, D. L., & Green, N. M. (1990). Structure of CaATPase: electron microscopy of frozen-hydrated crystals at 6 Å resolution in projection. J. Mol. Biol. 213, 529-538.

Tada, M. (1992). Molecular structure and function of phospholamban in regulating the calcium pump from sarcoplasmic reticulum. Ann. NY Acad. Sci. 671, 92-103.

Takeshima, H., Nishimura, S., Matsumoto, T., Ishida, H., Kangawa, K., Minamino, N., Matsuo, H., Ueda, M., Hanaoka, M., Hirose, T., & Numa, S. (1989). Primary structure and expression from complementary DNA of skeletal muscle ryanodine receptor. Nature 339, 439-445.

Tanabe, T., Beam, K. G., Powell, J. A., & Numa, S. (1988). Restoration of excitation-contraction coupling and slow calcium current in dysgenic muscle by dihydropyridine receptor complementary DNA. Nature 336, 134-139.

Tanabe, T., Mikami, A., Numa, S., & Beam, K. G. (1990a). Cardiac-type excitation-contraction coupling in dysgenic skeletal muscle injected with cardiac dihydropyridine receptor cDNA. Nature 344, 451-453.

Tanabe, T., Beam, K. G., Adams, B. A., Niidome, T., & Numa, S. (1990b). Regions of the skeletal muscle dihydropyridine receptor critical for excitation-contraction coupling. Nature 346, 567-569.

Tanabe, T., Adams, B. A., Numa, S., & Beam, K. G. (1991). Repeat I of the dihydropyridine receptor is critical in determining calcium channel activation kinetics. Nature 352, 800-803.

Taylor, K., Dux, L., & Martonosi, A. (1984). Structure of the vanadate-induced crystals of sarcoplasmic reticulum Ca^{2+}-ATPase. J. Mol. Biol. 174, 193-204.

Taylor, K. A., Dux, L., & Martonosi, A. (1986a). Three-dimensional reconstruction of negatively stained crystals of the Ca^{2+}-ATPase from muscle sarcoplasmic reticulum. J. Mol. Biol. 187, 417-427.

Taylor, K. A., Ho, M.-H., & Martonosi, A. (1986b). Image analysis of Ca^{2+}-ATPase from sarcoplasmic reticulum. Ann. NY Acad. Sci. 483, 31-43.

Taylor, K. A., Mullner, N., Pikula, S., Dux, L., Peracchia, C., Varga, S., & Martonosi, A. (1988). Electron microscope observations on Ca^{2+}-ATPase microcrystals in detergent-solubilized sarcoplasmic reticulum. J. Biol. Chem. 263, 5287-5294.

Wagenknecht, T., Grassucci, R., Frank, J., Saito, A., Inui, M., & Fleischer, S. (1989). Three dimensional architecture of the calcium channel/foot structure of sarcoplasmic reticulum. Nature 338, 167-170.

Weber, A. (1966). Energized calcium transport and relaxing factors. Curr. Top. Bioenerg. 1, 203-254.

Zorzato, F., Fujii, J., Otsu, K., Phillips, M., Green, N. M., Lai, F. A., Meissner, G., & MacLennan, D. H. (1990). Molecular cloning of cDNA encoding human and rabbit forms of the Ca^{2+} release channel (ryanodine receptor) of skeletal muscle sarcoplasmic reticulum. J. Biol. Chem. 265, 2244-2256.

Chapter 10

The Ribosome

RICHARD BRIMACOMBE

Principles of Medical Biology, Volume 2
Cellular Organelles, pages 253-273

ISBN:1-55938-803-X

INTRODUCTION

The flow of genetic information from DNA to RNA to protein is a universal process in all cellular organisms, and the final stage in this process-namely, the translation of a messenger RNA sequence into a protein molecule—takes place on the ribosome. Each consecutive triplet sequence of three bases on the mRNA corresponds to a single amino acid according to the genetic code, and the amino acids to be incorporated into the protein chain are brought to the ribosome in the form of covalent aminoacyl-tRNA complexes. A molecule of tRNA consists of about 80 nucleotides, which are folded into a rather rigid L-shape (Kim et al., 1973). The amino acid is attached to the short arm of the "L," whereas at the end of the long arm of the "L" there is a triplet anti-codon sequence, which is able to bind to a coding triplet on the mRNA by Watson-Crick base-pairing. There is at least one (and often two or three) tRNA species corresponding to each amino acid, and the tRNA's are charged with their appropriate amino acids by specific aminoacyl-tRNA synthetase enzymes. Thus, the amino acids arriving at the ribosome have already been effectively translated, i.e., fitted to a nucleic acid sequence, by the aminoacyl-tRNA synthetases, and the function of the ribosome is to align these charged aminoacyl-tRNA molecules in the correct order with respect to the mRNA sequence, via base-pairing between the codon triplets on the mRNA and the anticodon triplets on the tRNAs. Peptide bond formation between the consecutive amino acids then takes place on the ribosome by a simple transesterification reaction.

The study of the structure and function of the ribosome is a fascinating subject, which is clearly of central importance in molecular biology. From a medical point of view it is equally important, since a large number of antibiotics are known to act by interfering with specific ribosomal functions (see Nierhaus et al., 1988 for review). These antibiotics are in turn often useful tools for studying the various steps in the protein biosynthetic process. Although the details of protein biosynthesis vary considerably from one class of organism to another, the fundamental principles involved are universal, and in this article I will briefly outline the current status of our knowledge of ribosome structure and function, concentrating on the most widely-studied ribosomal species, namely that from the eubacterium *Escherichia coli.*

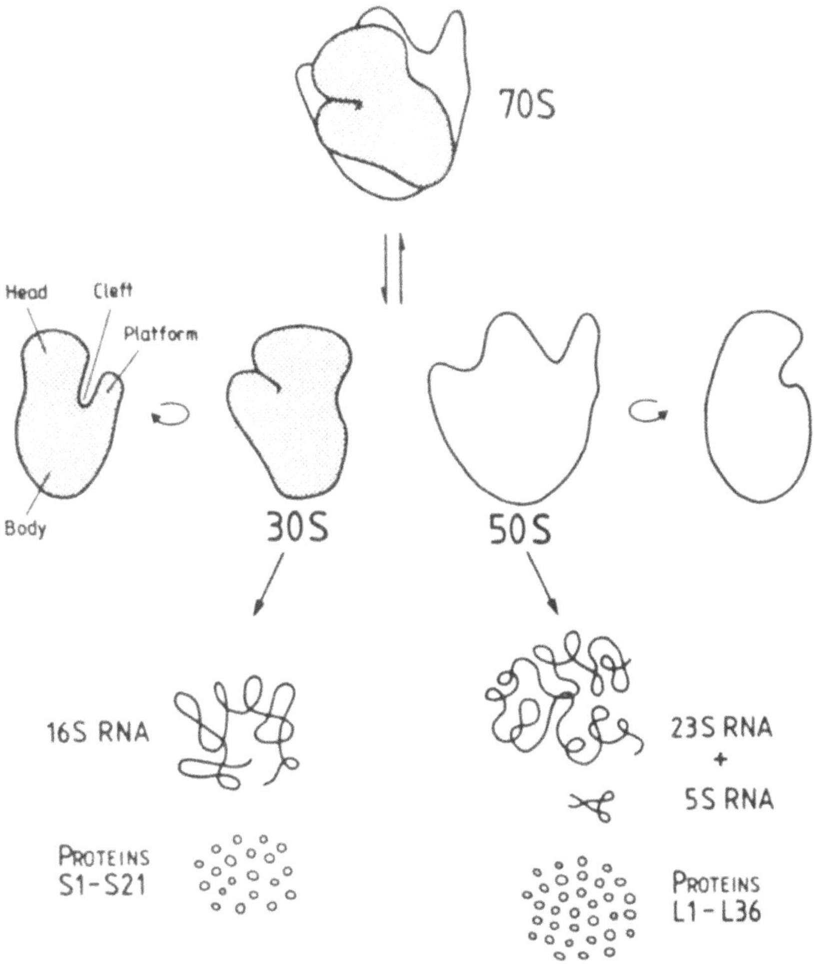

Figure 1. Components of the *E. coli* ribosome. The 70S ribosome (top) dissociates into 30S and 50S subunits, which are sketched in two orientations representing the forms seen by electron microscopy. Each subunit contains RNA and protein molecules as indicated. The S-value refers to the sedimentation constant of the particle concerned.

COMPONENTS OF THE *E. COLI* RIBOSOME

Every ribosome consists of two subunits of unequal size, and each subunit is itself a complex mixture of RNA and protein molecules

(Figure 1). In the *E. coli* ribosome (Wittmann, 1982), the smaller (30S) subunit contains a single (16S) RNA molecule, together with 21 different proteins. The proteins are named S1 to S21 according to their mobilities in a two-dimensional gel electrophoresis system. Correspondingly, the large (50S) subunit contains proteins named L1 to L36, together with two RNA molecules, one a large 23S RNA and the other a 5S RNA. The RNA comprises roughly two-thirds of the mass of each subunit, and the complete (70S) ribosome has a combined molecular weight of about 2×10^6. In addition, a number of protein factors are involved, which become transiently attached to the ribosome at various stages during the translation process.

FUNCTION

The synthesis of a protein on the ribosome (reviewed by Hershey, 1987) can be divided into three phases, namely the initiation step, the elongation process (during each cycle of which one amino acid is added to the growing polypeptide chain), and the termination step.

Initiation

In prokaryotes, the first step in the procedure is the recognition of a start signal on the mRNA, which interacts by base-pairing with a short sequence of three to six bases at the extreme 3′-terminus of the 16S ribosomal RNA in the 30S subunit. This start signal is a few bases upstream of the first coding triplet (usually AUG) on the mRNA, which is in turn recognized by a special initiator tRNA carrying the N-blocked amino acid formyl methionine. With the help of three protein initiation factors (named IF-1, IF-2, and IF-3) a 30S initiation complex is formed, consisting of the 30S ribosomal subunit and the formyl methionine-tRNA paired with the AUG codon on the mRNA. The 50S subunit is then added, forming the complete 70S ribosomal initiation complex, and the process of chain elongation can begin. At this stage the initiator tRNA is bound to a site on the ribosome which normally contains the tRNA attached to the growing polypeptide chain, and this site is accordingly known as the P-site.

Elongation

In each elongation cycle (Figure 2), the new aminoacyl-tRNA species corresponding to the next codon on the mRNA (the latter being read in

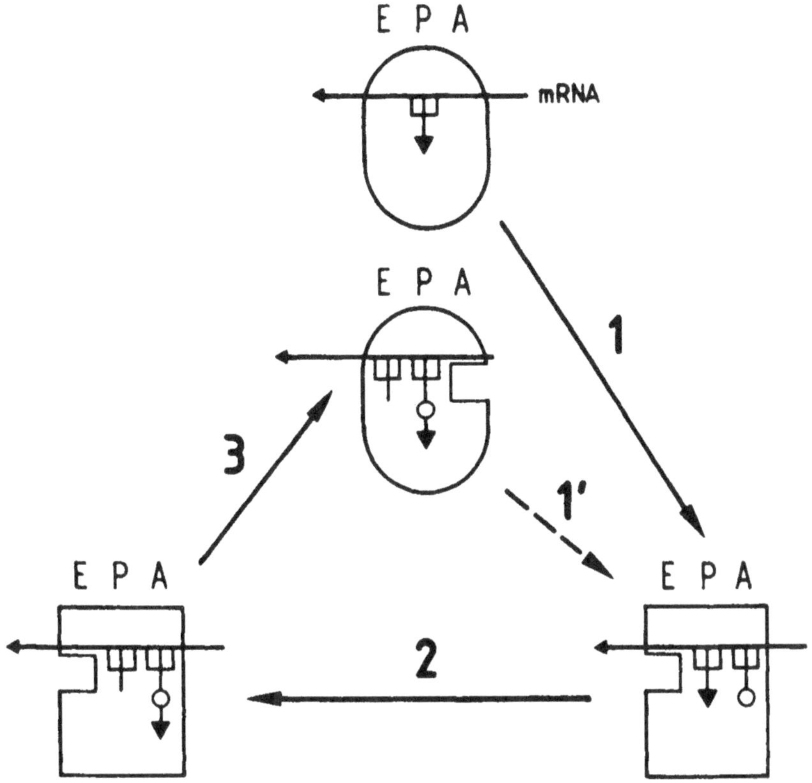

Figure 2. A simplified representation of the ribosomal elongation cycle. The 70S initiation complex is shown at the top, consisting of a ribosome, an mRNA molecule, and formyl methionine tRNA at the P-site; tRNA is denoted by the trident, and the methionyl residue by the triangle. In step 1, aminoacyl-tRNA (trident with open circle) is added to the A-site, giving a ribosome with low-affinity E-site (symbolized by the rectangular shape). In step 2, peptidyl transfer takes place. Step 3 is the translocation reaction, leading to a ribosome with empty tRNA at the E-site and peptidyl-tRNA at the P site. In this post-translocational mode the A-site has low affinity (symbolized by the return to the oval shape of the ribosome). In step 1´ the next elongation cycle begins, with loss of the empty tRNA from the E-site and concomitant arrival of the next aminoacyl-tRNA.

a 5´- to 3´-direction) is brought to the ribosome in the form of a ternary complex together with GTP and a protein elongation factor called EF-Tu. The aminoacyl-tRNA binds to a site on the ribosome known as the A-site (for aminoacyl or acceptor site), and this site is arranged in relation to the P-site in such a way that the short arms of each tRNA L-form lie close together. The formyl methionine residue (or in later cycles the polypeptide chain) is then transferred from the tRNA in the P-site to the newly added aminoacyl-tRNA at the A-site, thus giving rise to a peptidyl-tRNA molecule in which the peptide chain has been lengthened by one amino acid residue. This process is catalyzed by the peptidyl transferase, an enzymatic activity associated with the 50S subunit. At this stage the ribosomal P-site carries an empty tRNA molecule, whereas the peptidyl-tRNA is located at the A-site.

It should be noted that tRNA molecules are described by their amino acid specificity, and by their state of charging. Thus, $tRNA^{Phe}$ denotes an empty tRNA molecule, corresponding to the amino acid phenyl alanine. When this tRNA has been charged with phenyl alanine, the notation is Phe-$tRNA^{Phe}$.

The next stage in the elongation cycle is the all-important translocation step in which the whole complex (consisting of mRNA, the peptidyl-tRNA, and the empty tRNA) has to be shifted across the ribosome by one codon length, so as to enable the next elongation cycle to take place. This process requires another protein factor (elongation factor EF-G) together with GTP, and very probably involves the active participation of the ribosome itself. After translocation, the peptidyl-tRNA is again in the P-site, and the A-site is free to accept the next incoming aminoacyl-tRNA molecule.

The A- and P-sites can be operationally defined with the help of the antibiotic puromycin, which is an analogue of aminoacyl-tRNA. Puromycin binds to the A-site and can accept the peptidyl residue from a P-site-bound tRNA. Further elongation of the peptide chain is, however, prevented. Thus, a positive reaction with puromycin defines P-site binding of the ligand under study, whereas a negative reaction indicates A-site binding.

In every textbook that has been written during the last twenty years, the ribosomal cycle is described simply in terms of these two tRNA sites, the A- and P-sites. This description is, however, now known to be incomplete, because there is in fact a third tRNA site, known as the E-site (exit site, Nierhaus and Rheinberger, 1984). The presence of a third site could be demonstrated by the observation that under certain conditions three tRNA molecules can bind simultaneously to the ribosome, but the most convincing proof for the existence of this site comes from the following consideration. In the old two-site model, a ribosome in the pre-translocational state contains, as described above, the peptidyl-tRNA at the A-site, and an empty tRNA at the P-site. When the peptidyl-tRNA is now translocated to the P-site, the empty tRNA (having nowhere else to go in the two-site model) should concomitantly leave the ribosome. In fact this does not happen, and the empty tRNA can be shown to remain firmly bound to the ribosome after the translocation process. The empty tRNA is thus located at the third site, the E-site, and only leaves the ribosome when the next aminoacyl-tRNA arrives at the A-site (Figure 2). The existence of the third tRNA site has been demonstrated in archaebacteria and in eukaryotes as well as in *E. coli.*

Termination

During the elongation process, several ribosomes may be working on the same mRNA molecule, giving rise to the well-documented appearance of polysomes. As each successive ribosome reaches the end of the coding sequence on the mRNA, chain termination is signaled by one of the stop codons, UAA, UAG or UGA, and the completed polypeptide chain is released from the ribosome with the help of one of two release factors, RF-1 or RF-2. The 70S ribosome dissociates into 30S and 50S subunits, and a new cycle of protein synthesis can begin.

STRUCTURE

Any deeper understanding of the molecular mechanisms involved in the functional processes just described obviously requires a detailed knowledge of the ribosome structure, and many different methods have been applied in a wide variety of structural studies (see Wittmann, 1983 for review). These cover the structure and properties of the individual ribosomal protein and RNA molecules, the interactions between RNA and proteins, the three-dimensional arrangement of the various components in the intact ribosomal subunits, and the properties of the ribosomes as a whole. Only a selection of the most important methods will be described in the following sections.

Electron Microscopy

The 30S and 50S subunits of the *E. coli* ribosome have approximate dimensions of 220 x 140 x 110 Å and 230 x 230 x 150 Å, respectively, as determined by electron microscopy. More importantly each subunit has a characteristic and recognizable shape (Figure 1), which enables ribosomal components or ligands to be localized on the subunit surface, by the technique of immune electron microscopy (Stöffler and Stöffler-Meilicke, 1986; Oakes et al., 1986). Thus, the 30S subunit consists of a head, a body and a platform, whereas the 50S subunit has three distinct protuberances on its upper side. Antibodies against the ribosomal component or ligand of interest can be prepared, and the points of attachment of the antibodies to the subunit surface in relation to these characteristic features can be observed in the electron microscope. In this way the positions of epitopes in many ribosomal proteins, as well as a number of modified bases in the ribosomal RNA (the latter summarized in Gornicki et al., 1984) have been localized on the 30S or 50S subunits. All of these studies were made using negatively stained ribosomal particles for the electron microscopy. More recently, a three-dimensional reconstruction of the *E. coli* 70S ribosome has been reported (Frank et al. 1991) from electron micrographs in amorphous ice (i.e., without staining); this method provides a more detailed description of the ribosomal morphology, and shows several new features (see section on Structure-Function Correlation below).

Neutron Scattering

A more quantitative description of the spatial arrangement of the ribosomal proteins has been provided by low-angle neutron scattering. This method relies for its success on the important property of the *E. coli* ribosome that biologically active 30S and 50S ribosomal subunits can be reconstituted from their isolated protein and RNA moieties. Thus, subunits can be prepared by reconstitution in which one or more components have been modified to suit the needs of a particular experiment. For neutron scattering studies with the 30S subunit, selected pairs of deuterated proteins are incorporated into the subunit, and the neutron scattering technique allows the distance between the mass centers of the two deuterated proteins to be determined, since deuterium has markedly different neutron scattering properties from those of hydrogen. As a result of a large number of such measurements carried out over the last fifteen years, the relative positions of the mass centers of all 21 proteins of the 30S subunit have now been triangulated (Capel et al., 1988). A similar series of experiments (but using reconstituted subunits with two normally hydrogenated proteins in an otherwise deuterated subunit) is currently in progress with the 50S subunit (May et al., 1992).

Protein-Protein Cross-Linking, and Amino Acid Sequences

Pairs of proteins that are close neighbors in the ribosome can also be investigated by cross-linking techniques. The cross-linked protein pairs are isolated from ribosomal subunits that have been treated with a suitable bifunctional reagent, and the components in each cross-linked pair can be identified either immunologically (Walleczek et al., 1989), or (if a reversible cross-linking reagent has been used) by gel electrophoresis (e.g., Lambert et al., 1983). Taken together with the immune electron microscopic and neutron scattering data just mentioned, a coherent picture of the three-dimensional arrangement of the proteins in the *E. coli* 30S subunit has emerged, and the corresponding arrangement of the proteins in the 50S subunits is also becoming clear (Walleczek et al, 1988).

In a few cases, the cross-linking studies have been pursued to a more intimate level of resolution, with the identification of the amino acids involved in the cross-link (e.g., Brockmöller and Kamp, 1986). Such an analysis of course requires knowledge of the amino acid sequences of the proteins concerned, and these sequences have been determined for all of the *E. coli* ribosomal proteins, as well as for a large number of ribosomal proteins from a variety of other organisms (Wittmann-Liebold, 1986). The sequence data furthermore provide a useful framework for studying the evolutionary relationships between the different organisms concerned (Wittmann-Liebold et al., 1990).

RNA Primary Sequences and Secondary Structure

As with the ribosomal proteins, a large number of sequences of ribosomal RNA molecules are now available. These sequences have for the most part been determined via the corresponding ribosomal DNA, and now total over 700 published 5S RNA sequences from different species (Specht et al., 1991), over 920 16S-type (or small subunit RNA) sequences (De Rijk et al., 1992) and over 120 large subunit RNA sequences (Gutell et al., 1992). The length of these ribosomal RNA molecules varies considerably from one class of organism to another. The small subunit RNA is only ca 640 nucleotides long in the mitochondria from trypanosomes, whereas in bacteria it contains ca 1500 nucleotides (1542 in *E. coli*), and in cytoplasmic ribosomes from higher eukaryotes it reaches a length of ca 1870 nucleotides. The corresponding approximate lengths for the large subunit RNA molecules are 1230, 2900

(2904 in *E. coli*), and 4800 nucleotides, respectively. In contrast, the 5S RNA is rather constant in length (120 nucleotides), but in some species of small mitochondrial ribosome this RNA species is lacking altogether.

Unquestionably the most important result to have emerged from all the sequence data has been the derivation of reliable general secondary structure models for the ribosomal RNA molecules. RNA chains are able to fold back on themselves, forming complex double-helical hairpin-loop structures which are held together by classical Watson-Crick base-pairing. Putative secondary structural features of this nature can be tested directly by comparing the sequences of pairs of ribosomal RNA molecules from related organisms, since—if the postulated secondary structure element is correct—then base changes between the two sequences being compared must compensate one another in the double-helical region (e.g., an A-U pair in one sequence becoming a G-C pair in the other), in order to maintain the integrity of the base-paired double helix concerned (Fox and Woese, 1975). The current versions of the secondary structures of the *E. coli* 16S and 23S RNA molecules are based on literally tens of thousands of such compensating base changes (see e.g., Noller, 1984; Brimacombe and Stiege, 1985), and the structures are also supported by data from a variety of experimental techniques, such as accessibility to single- or double-strand specific nucleases, susceptibility to modification by base-specific chemical reagents, or isolation of short base-paired RNA fragments by two-dimensional gel electrophoresis (see Ehresmann et al., 1987 for review).

The secondary structure models furthermore indicate clearly that (a) there is a conserved core of secondary structure in all classes of ribosomal RNA molecules, (b) the differences in length already mentioned between the RNA molecules from different evolutionary classes are accommodated by large deletions or insertions (relative to *E. coli*) in discrete parts of the structures, and (c) other odd species of ribosomal RNA associated with the large subunit from higher organisms, such as 5.8S, 4.5S or 2S RNA, all have clear counterparts within the eubacterial 23S RNA (see Brimacombe and Stiege, 1985).

Intra-RNA and RNA-Protein Cross-Linking

The *E. coli* 16S and 23S RNA secondary structures provide a basis both for the further folding of the RNA molecules into compact three-dimensional structures, and also for relating the spatial arrangement of the RNA to that of the ribosomal proteins as determined by the techniques discussed above. For these purposes, the most direct information comes from the application of intra-RNA and RNA-protein cross-linking methods, respectively. The cross-links are introduced into intact ribosomal subunits or 70S ribosomes by treatment with appropriate bifunctional reagents, the RNA is subjected to some kind of partial nuclease digestion procedure, and the cross-linked complexes—consisting either of two fragments of RNA linked together, or an RNA fragment linked to a ribosomal protein—are isolated by gel electrophoresis. The cross-linked sites on the RNA are analyzed by sequencing procedures, and the cross-linked proteins are identified immunologically. In this way a network of neighborhoods or contact points between different pairs of sites on the RNA, or between sites on the RNA and various individual ribosomal proteins, has been accumulated for both the 30S and 50S subunits (Brimacombe et al., 1988, 1990).

Foot-Printing

An alternative approach to RNA-protein cross-linking is the foot-printing method. Here, the patterns of accessibility of the RNA or a ribosomal subunit to chemical modification or enzymatic digestion are compared in the presence or absence of a particular ribosomal protein, with the help of the reconstitution procedures already mentioned. The sites on the RNA that have been attacked can be rapidly scanned using reverse transcriptase in conjunction with a series of deoxyoligonucleotide primers set at strategic intervals along the RNA molecule, and a comparison of the modification or digestion patterns in the presence or absence of the protein under study indicates which bases are protected by that protein (Stern et al., 1988). The method has also been usefully applied to ligands such as tRNA or antibiotics (see below), but it does have the notable disadvantage that direct protection effects cannot be distinguished from allosteric or long-range protection effects due to structural rearrangements caused by binding of the protein or ligand concerned.

Three-Dimensional Models of the Ribosomal RNA

By combining all the data from the methods discussed above, models have been derived for the three-dimensional distribution of the ribosomal components in both the 30S subunit (Brimacombe et al, 1988; Stern et al., 1988) and the 50S subunit (Mitchell et al., 1990). These models essentially describe the arrangement of the double-helical segments of the ribosomal RNA, the helices being constrained (a) by their interconnections within the primary sequence of the RNA, (b) by contacts found from intra-RNA cross-linking studies, (c) by the positions of individual bases on the subunit surface found by immune electron microscopy, and (d) by their relationship to the arrangement of the ribosomal proteins, as revealed by RNA- protein cross-linking or foot-printing experiments.

A computer graphics version of the 30S model derived in this laboratory is shown in Figure 3. The model has the recognizable shape of the 30S subunit (cf. Figure 1), and its dimensions agree with those found by electron microscopy. Furthermore, the model shows that the large variations in the length of the ribosomal RNA molecules from different classes of organisms are not only accommodated into discrete regions of the secondary structures (as already discussed above), but that these regions are also grouped into distinct three-dimensional domains at the upper and lower extremities of the 30S subunit. In contrast, most of the conserved core of the secondary structure is concentrated into a broad band around the middle of the subunit (Figure 3).

A corresponding situation can be observed with the 50S subunit. In this case, the conserved core of secondary structure covers the interface side of the subunit (i.e., the side which interacts with the 30S subunit), and the regions where insertions or deletions in the RNA occur in other organisms are located on the opposite side of the model (Mitchell et al., 1990). Many more features of both the 30S and 50S models concur with various other sets of experimental data (discussed in Brimacombe et al., 1988; Mitchell et al., 1990), although more recent findings indicate that some regions of the structures need to be revised (see below). This type of discrepancy precisely illustrates the importance of these model-building studies, since although the models are still very crude (and in no way comparable to crystallographically-derived atomic models), they provide a framework for incorporating all of the available structural information into a single concept, as well as for correlating the complex structure of the ribosome with its equally complex function.

STRUCTURE-FUNCTION CORRELATION

Location of Functionally Important Sites in the 30S and 50S Subunits

Electron microscopic studies have shown that the decoding site on the 30S subunit (that is to say the region where the interaction between

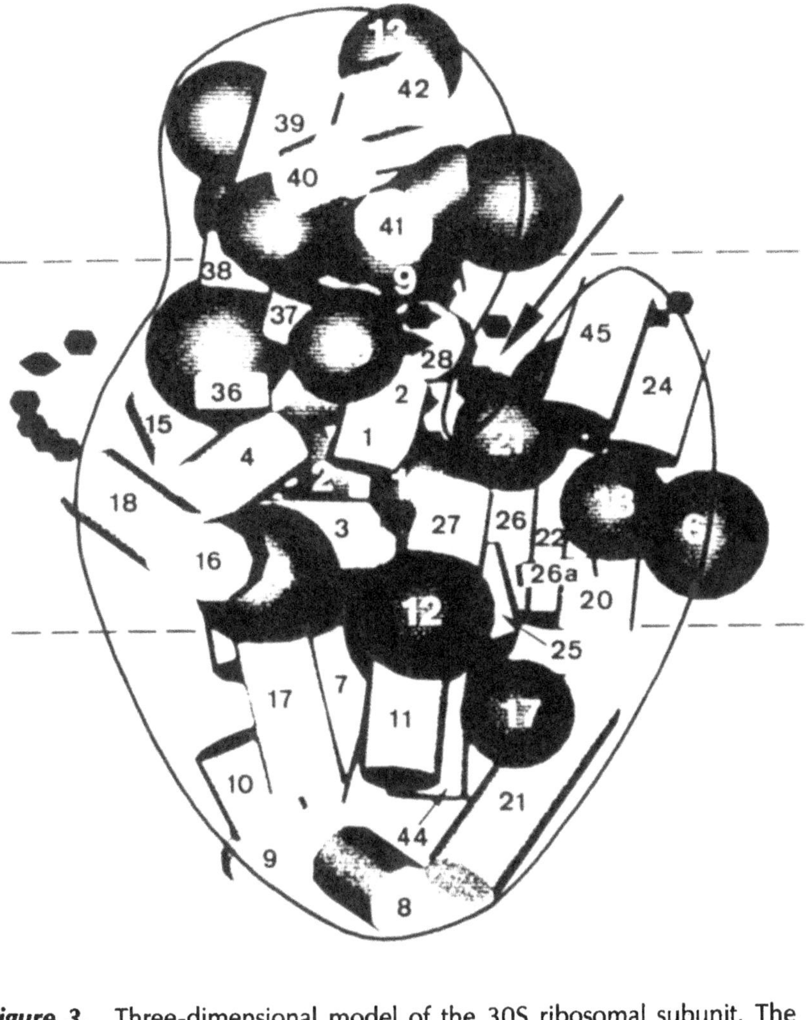

Figure 3. Three-dimensional model of the 30S ribosomal subunit. The spheres represent the ribosomal proteins, and the cylinders the double-helical regions in the 16S RNA, numbered as in Brimacombe et al (1988). Single-stranded regions of the RNA are not included. The small polygons represent functionally important sites in the RNA (see Brimacombe, 1988, for review). The model is enclosed in a rough silhouette representing the electron microscopically observed shape of the 30S subunit (cf. Fig. 1), and the broad arrow indicates the supposed site of codon-anticodon interaction. The dotted lines encompass the region of the subunit in which the majority of the conserved core of the RNA structure is found (see text).

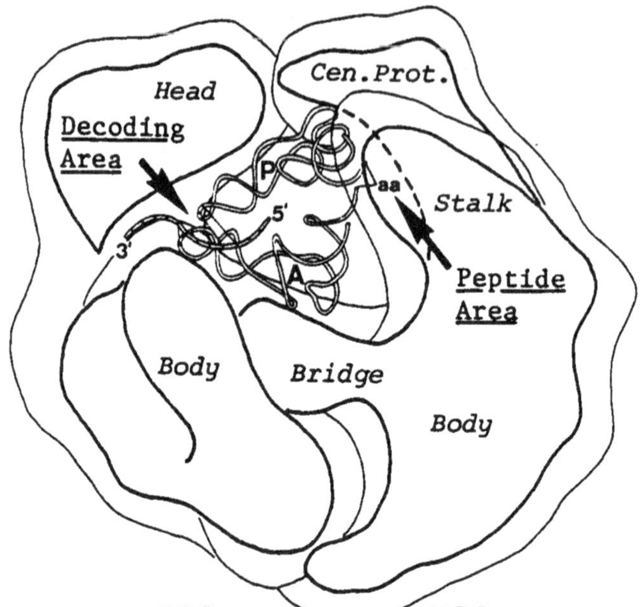

Figure 4. Sketch of the electron microscopic model of the 70S ribosome (Frank et al., 1991) showing how the tRNA-mRNA complex could be accomodated. The mRNA is marked by its 5'- and 3'-ends, and the A- and P-site tRNA positions are indicated, aa denoting the aminoacyl or peptide moiety. The orientation of the 70S ribosome is at 90° to that shown in Figure 1 (top), Cen. Prot. indicating the central protuberance of the 50S subunit, and the stalk corresponding to the right-hand of the three 50S protuberances as depicted in Figure 1.

the codons on the mRNA and the anticodons on the tRNA takes place) is located in or around the base of the cleft separating the head of the 30S subunit from the platform (see e.g., Gornicki et al., 1984; Figure 3). Correspondingly, in the 50S subunit, the peptidyl transferase center—the region where the successive amino acids become linked together to form the nascent peptide chain—is located at the base of the central protuberance of the subunit (see Stöffler and Stöffler-Meilicke, 1986). In the more recent electron microscopic study of Frank et al. (1991), already mentioned above, the platform of the 30S subunit is more flattened and less pronounced, and a distinct bridge can be seen between the 30S and 50S subunits. In consequence, there is a well-defined interface cavity between the two subunits, which has dimensions which correspond well with those of the tRNA-mRNA complex. A sketch of this model, showing one possible arrangement of the A- and P-site tRNAs, together with a segment of mRNA, is illustrated in Figure 4.

It can be seen from Figure 4 that the two ends of the two L-shaped tRNA molecules are located close to one another both in the decoding area on the 30S subunit (thus enabling the P- and A-site anticodons to interact with adjacent codons on the mRNA) and in the peptide area on the 50S subunit (thus enabling the peptidyl transfer to take place). On the other hand, a large number of positions in both the 16S and 23S RNA molecules have been implicated as being important for these functions; these positions include foot-printing sites of A- and P-site bound tRNA (Moazed and Noller, 1989, 1990), foot- printing sites of various antibiotics (e.g., Moazed and Noller, 1987), as well as sites of mutation causing resistance to antibiotics (discussed in Moazed and Noller, 1987, 1989, 1990). These positions are widely distributed throughout the primary and secondary structures of the 16S and 23S RNA, but nevertheless inspection of the models shows that in three dimensions all the sites are clustered into groups. In the case of the 23S RNA, the functional sites form a single cluster at the base of the central protuberance of the 50S subunit (Mitchell et al., 1990; cf. Figure 4), whereas in the case of the 16S RNA there appear to be two groups of sites on opposite sides of the 30S subunit (Brimacombe et al., 1988; Stern et al, 1988; see Fig. 3). More recent data—from Site-Directed Cross-Linking studies (below) suggests that this latter conclusion is incorrect and reflects errors in the model-building.

Site-Directed Cross-Linking With Functional Ligands

The foot-printing technique for locating the binding sites on the ribosome of ligands such as tRNA has two disadvantages. One (already noted) is the inability to discriminate between direct and allosteric effects, and the other is that the technique does not have the potential of defining the orientation of the ligand concerned. A more sophisticated technique, which is increasingly applied to the study of tRNA and mRNA binding to the ribosome, is the site-directed cross-linking method; this is in effect an extension of the affinity labeling approach. Here, with the help of modern methods for manipulating or synthesizing nucleic acids, tRNA or mRNA analogues can be made which

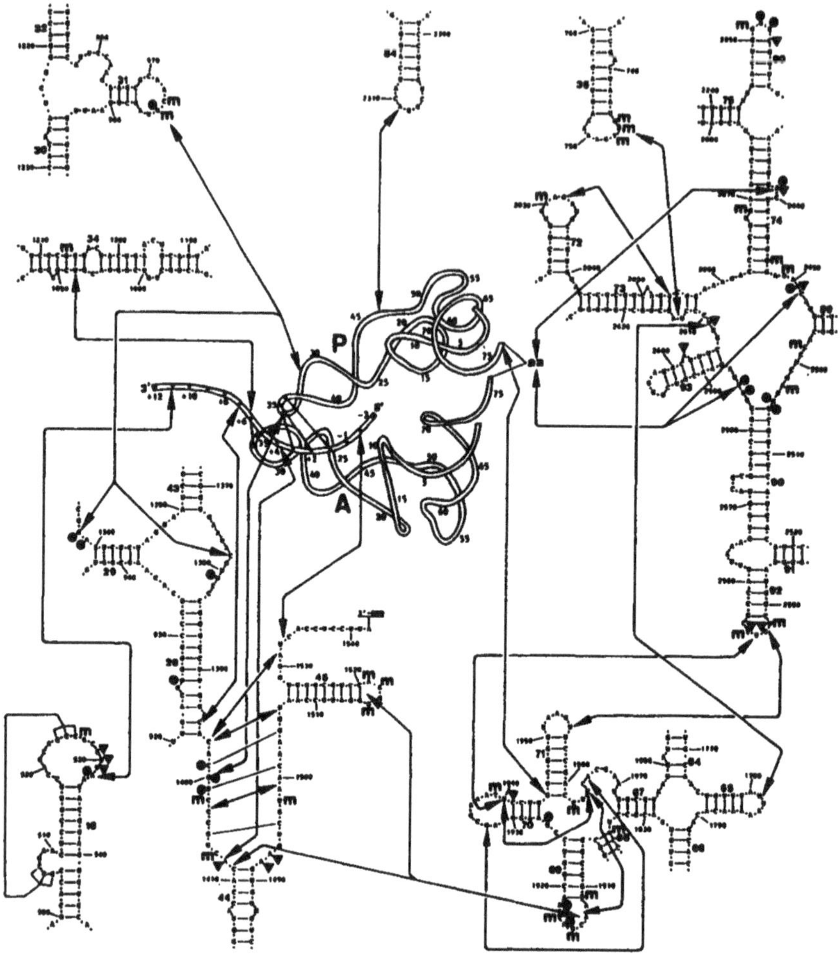

Figure 5. The tRNA-mRNA complex (as depicted in Figure 4), showing positions of site-directed cross-linking to regions of the 16S or 23S RNA, together with other relevant cross-linking data in these functionally important areas (which represent about 15% of the total ribosomal RNA). The figure is a modified version of that in Brimacombe et al. (1993); see the latter publication for references to the individual cross-linking data sets. The RNA elements on the left are from 16S RNA (numbered as in Brimacombe et al., 1988), those on the right from 23S RNA (numbered as in Mitchell et al., 1990). The closed circles and triangles represent sites of foot-printing from P-site or A-site bound tRNA, respectively (Moazed and Noller, 1989, 1990), and m denotes modified nucleotides in the ribosomal RNA.

carry a modified nucleotide (e.g., with a photo-reactive group) at any desired position. These molecules can be bound to the ribosome under physiological conditions, and the ribosomal components which are able to react with the modified nucleotide can be identified and the precise sites of cross-linking to the latter analysed. The method yields data relating both to the ribosomal proteins (e.g., Ofengand et al., 1986) and to the ribosomal RNA, the latest status of the results in the case of the RNA being summarized in Figure 5.

Figure 5 shows (on the left-hand side) the locations in 16S RNA of site-directed cross-links from the anticodon regions of A-site and P-site bound tRNA, and from mRNA. The cross-links correlate closely with the positions of tRNA foot-print sites and also to the modified nucleotides in the 16S RNA (Brimacombe et al., 1993). In particular, the cross-links from the 3'-region of the mRNA indicate that the RNA regions containing the clusters of functional sites which appeared to be separated in the 30S subunit (Figure 3) must in fact lie close together, since they encompass only a very short stretch (ca 7 nucleotides) of the mRNA (Dontsova et al., 1992). Similarly, the right-hand side of Figure 5 gives the locations in 23S RNA of site-directed cross-links from the elbow and amino-acyl regions of tRNA (see Brimacombe et al., 1993 for references), again indicating a close correlation with the positions of tRNA foot-print sites and modified nucleotides, as well as a clustering of several functionally important regions of the 23S molecule that are widely separated in the primary and secondary structure. Figure 5 only illustrates about 15% of the total ribosomal RNA, and in fact many other regions of the 16S and 23S molecules are connected to these regions by a network of further cross-links as well as by the secondary structure of the RNA itself (see Mitchell et al., 1990). Bearing in mind the wealth of data already mentioned relating to RNA-protein neighborhoods, it should be clear that the constraints on the three-dimensional folding of the functional regions of the RNA within the ribosome are enormously complex; the challenge is now to see how far these constraints, together with the new results that are continually appearing, can be accommodated into self-consistent models for the active centers of the 30S and 50S subunits.

CONCLUSIONS AND PROSPECTS

As I hope to have shown in this brief survey, the exciting feature of ribosome research at the present time is that a whole range of different experimental approaches relevant to both structure and function have, after many years of non-interaction or even divergence, begun to rapidly converge together. The progress will not stop here, and, in conclusion, two areas of research should be mentioned which are likely to have a considerable impact on the further development of the subject: Site-directed mutagenesis and crystallization of ribosomal proteins and whole ribosomal subunits.

Site-Directed Mutagenesis

Again with the help of modern techniques for manipulating nucleic acids, it is now possible to generate ribosomal RNA molecules carrying mutations, insertions or deletions at any desired position. Many research groups are engaged in applying these techniques (first used in the ribosome by Zwieb and Dahlberg, 1984), with a view to investigating the functional importance of individual nucleotides or regions of the ribosomal RNA. The number of possibilities for alterations of this nature is of course astronomical, but the list of useful targets for mutagenesis should become clearer concomitantly with our increasing understanding of the subunit structures.

Crystallization of Ribosomal Proteins and Whole Ribosomal Subunits

X-ray crystallography remains the ultimate tool in structural determinations, and the crystallization and analysis of individual ribosomal proteins has begun to gather momentum (Ramakrishnan and White, 1992). Furthermore, crystals of intact ribosomal subunits, in particular those from halophilic or thermophilic bacteria (Müssig et al., 1989), are now being obtained with increasingly high quality. Thus a crystallographic analysis of complete ribosomal subunits at atomic resolution no longer appears to be such a remote possibility. Nevertheless, the problem is still one of daunting magnitude, and the crystallographers will need to make maximum use of all the information from the lower-resolution studies summarized in this article.

REFERENCES

Brimacombe, R. (1988). The emerging three-dimensional structure and function of 16S ribosomal RNA. Biochemistry 27, 4207-4214.

Brimacombe, R., Atmadja, J., Stiege, W., & Schüler, D., (1988). A detailed model of the three-dimensional structure of *E. coli* 16S ribosomal RNA *in situ* in the 30S subunit. J. Mol. Biol. 199, 115-136.

Brimacombe, R., Gornicki, P., Greuer, B., Mitchell, P., Osswald, M., Rinke-Appel, J., Schüler, D., & Stade, K. (1990). The three-dimensional structure and function of *E. coli* ribosomal RNA, as studied by cross-linking techniques. Biochim. Biophys. Acta 1050, 8-13.

Brimacombe, R., Mitchell, P., Osswald, M., Stade, K., & Bochkariov, D. (1993). Clustering of modified nucleotides at the functional center of bacterial ribosomal RNA. FASEB J. 7, 161-167.

Brimacombe, R. & Stiege, W. (1985). Structure and function of ribosomal RNA. Biochem. J. 229, 1-17.

Brockmöller, J. & Kamp, R.M. (1986). Isolation and identification of two protein-protein cross-links within the two subunits of B. *stearothermophilus* ribosomes. Biol. Chem. Hoppe-Seyler 367, 925-935.

Capel, M.S., Kjeldgaard, M., Engelman, D.M. & Moore, P.B. (1988). Positions of S2, S13, S16, S17, S19 and S21 in the 30S ribosomal subunit of *E. coli*. J. Mol. Biol. 200, 65-87.

De Rijk, P., Neefs, J.M., Van de Peer, Y., & De Wachter, R. (1992). Compilation of small ribosomal subunit RNA sequences. Nucleic Acids Res. Suppl. 20, 2075-2089.

Dontsova, O., Dokudovskaya, S., Kopylov, A., Bogdanov, A., Rinke-Appel, J., Jünke, N. & Brimacombe, R. (1992). Three widely separated positions in the 16S RNA lie in or close to the ribosomal decoding region; a site-directed cross-linking study with mRNA analogues. EMBO J. 11, 3105-3116.

Ehresmann, C., Baudin, F., Mougel, M., Romby, P., Ebel, J.P., & Ehresmann, B. (1987). Probing the structure of RNAs in solution. Nucleic Acids Res. 15, 9109-9128.

Fox, G. & Woese, C.R. (1975). 5S RNA secondary structure. Nature 256, 505-507.

Frank, J., Penczek, P., Grassucci, R., & Srivastava, S. (1991). Three-dimensional reconstruction of the 70S *E. coli* ribosome in ice; the distribution of ribosomal RNA. J. Cell Biol. 115, 597-605.

Gornicki, P., Nurse, K., Hellmann, W., Boublik, M. & Ofengand, J. (1984). High resolution localization of the tRNA anticodon interaction site on the *E. coli* 30S ribosomal subunit. J. Biol. Chem. 259, 10493-10498.

Gutell, R.R., Schnare, M.N., & Gray, M.W. (1992). Compilation of large subunit (23S and 23S-like) ribosomal RNA structures. Nucleic Acids Res. Suppl. 20, 2095-2109.

Hershey, J.W.B. (1987). Protein synthesis. In: *Escherichia coli* and *Salmonella typhimurium* (Neidhardt, F.C., Ed.), pp. 613-647, American Society for Microbiology, Washington DC.

Kim, S.H., Quigley, G.J., Suddath, F.L., McPherson, A., Sneden, D., Kim, J.J., Weinzierl, J., & Rich, A. (1973). Three-dimensional structure of yeast phenylalanine tRNA: Folding of the polynucleotide chain. Science 179, 285-288.

Lambert, J.M., Boileau, G., Cover, J.A., & Traut, R.R. (1983). Cross-links between ribosomal proteins of 30S subunits in 70S tight couples and in 30S subunits. Biochemistry 22, 3913-3920.

May, R.P., Nowotny, V., Nowotny, P., Voss, H., & Nierhaus, K.H. (1992). Inter-protein distances within the large subunit from *E. coli* ribosomes. EMBO J. 11, 373-378.

Mitchell, P., Osswald, M., Schüler, D., & Brimacombe, R. (1990). Selective isolation and detailed analysis of intra-RNA cross-links induced in the large ribosomal subunit of *E. coli;* a model for the tertiary structure of the tRNA binding domain in 23S RNA. Nucleic Acids Res. 18, 4325-4333.

Moazed, D. & Noller, H.F. (1987). Interaction of antibiotics with functional sites in 16S ribosomal RNA. Nature 327, 389-394.

Moazed, D. & Noller, H.F. (1987). Interaction of tRNA with 23S RNA in the ribosomal A, P and E sites. Cell 57, 585-597.

Moazed, D. & Noller, H.F. (1990). Binding of tRNA to the ribosomal A and P sites protects two distinct sets of nucleotides in 16S rRNA. J. Mol. Biol. 211, 135-145.

Müssig, J., Makowski, I., Von Böhlen, K., Hansen, H., Bartels, K.S., Wittmann, H.G., & Yonath, A. (1989). Crystals of wild-type, mutated, derivatized and complexed 50S ribosomal subunits from *B. stearothermophilus* suitable for X-ray analysis. J. Mol. Biol. 205, 619-621.

Nierhaus, K.H., Brimacombe, R., & Wittmann, H.G. (1988). Inhibition of protein synthesis by antibiotics. In: Perspectives in Anti-Infective Therapy (Jackson, G.G., Schlumberger, H.D., & Zeiler,H.J., Eds.), pp. 29-40, Friedrich Vieweg & Sohn, Braunschweig/Wiesbaden, Germany.

Nierhaus, K.H. & Rheinberger, H.J. (1984). An alternative model for the elongation cycle of protein biosynthesis. Trends in Biochem. Sciences 9, 428-432.

Noller, H.F. (1984). Structure of ribosomal RNA. Ann. Rev. Biochem. 53, 119-162.

Oakes, M., Henderson, E., Scheinman, A., Clark, M., & Lake, J.A. (1986). Ribosome structure, function and evolution: Mapping ribosomal RNA, proteins and functional sites in three dimensions. In: Structure, Function and Genetics of Ribosomes (Hardesty, B. & Kramer, G., Eds.), pp. 47-67, Springer-Verlag, Heidelberg & New York.

Ofengand, J., Ciesiolka, J., Denman, R., & Nurse, K. (1986). Structural and functional interactions of the tRNA-ribosome complex. In: Structure, Function and Genetics of Ribosomes (Hardesty, B. & Kramer, G., Eds.), pp. 473-494, Springer-Verlag, Heidelberg & New York.

Ramakrishnan, V. & White, S.W. (1992). The structure of ribosomal protein S5 reveals sites of interaction with 16S rRNA. Nature 358, 768-771.

Specht, T., Wolters, J., & Erdmann, V.A. (1991). Compilation of 5S rRNA and 5S rRNA gene sequences. Nucleic Acids Res. Suppl. 19, 2189-2191.

Stern, S., Weiser, B., & Noller, H.F. (1988). Model for the three-dimensional folding of 16S ribosomal RNA. J. Mol. Biol. 204, 447-481.

Stöffler, G. & Stöffler-Meilicke, M. (1986). Immuno-electron microscopy on *E. coli* ribosomes. In: Structure, Function and Genetics of Ribosomes (Hardesty, B. & Kramer, G., Eds.), pp. 28-46, Springer-Verlag, Heidelberg & New York.

Walleczek, J., Martin, T., Redl, B., Stöffler-Meilicke, M., & Stöffler, G. (1989). Comparative cross-linking study on the 50S ribosomal subunit from *E. coli.* Biochemistry 28, 4099-4105.

Walleczek, J., Schüler, D., Stöffler-Meilicke, M., Brimacombe, R., & Stöffler, G. (1988). A model for the spatial arrangement of the proteins in the large subunit of the *E. coli* ribosome. EMBO J. 7, 3571-3576.

Wittmann, H.G. (1982). Components of bacterial ribosomes. Ann. Rev. Biochem. 51, 155-183.

Wittmann, H.G. (1983). Architecture of prokaryotic ribosomes. Ann. Rev. Biochem. 52, 35-65.

Wittmann-Liebold, B. (1986). Ribosomal proteins. Their structure and evolution. In: Structure, Function and Genetics of Ribosomes (Hardesty, B. & Kramer, G., Eds.), pp. 326-361, Springer-Verlag, Heidelberg & New York.

Wittmann-Liebold, B., Köpke, A., Arndt, E., Kròmer, W., Hatakeyama, T., & Wittmann, H.G. (1990). Sequence comparison and evolution of ribosomal proteins and their genes. In: The Ribosome, Structure, Function & Evolution (Hill, W.E., Dahlberg, A.E., Garrett, R.A., Moore, P.B., Schlessinger, D., & Warner, J.R., Eds.), pp. 598-616, ASM Press, Washington D.C.

Zwieb, C. & Dahlberg, A.E. (1984). Point mutations in the middle of 16S ribosomal RNA of *E. coli* produced by deletion loop mutagenesis. Nucleic Acids Res. 12, 4361-4375.

NOTE ADDED IN PROOF

For a more recent description of the status of the topographical data on ribosomal RNA see Brimacombe, R. (1995). The structure of ribosomal RNA; a three-dimensional jigsaw puzzle. Eur. J. Biochem., in press.

INDEX

www.ingramcontent.com/pod product-compliance
Lightning Source LLC
La Vergne TN
LVHW011748200926
848484LV00003B/651
* 9 7 8 1 5 5 9 3 8 8 0 3 0 *